Mathematical Models of Hysteresis

I.D. Mayergoyz

Mathematical Models of Hysteresis

With 167 Illustrations

Springer-Verlag
New York Berlin Heidelberg London
Paris Tokyo Hong Kong Barcelona

I.D. Mayergoyz
Electrical Engineering Department
Institute for Advanced Computer Studies
University of Maryland
College Park, Maryland 20742

Library of Congress Cataloging-in-Publication Data

Mayergoyz, I.D.
Mathematical models of hysteresis / I.D. Mayergoyz.
p. cm.
Includes index.
ISBN 0-387-97352-4 (alk. paper)
1. Hysteresis—Mathematical models. I. Title.
QC754.2.H9M38 1991
538′.3—dc20 90-39253
CIP

Printed on acid-free paper

Typeset by Asco Trade Typesetting Ltd., Quarry Bay, Hong Kong.
Printed and bound by BookCrafters, Chelsea, Michigan.
Printed in the United States of America.

9 8 7 6 5 4 3 2 1

ISBN 0-387-97352-4 Springer-Verlag New York Berlin Heidelberg
ISBN 3-540-97352-4 Springer-Verlag Berlin Heidelberg New York

To my children
Sasha and Anna

Preface

The purpose of this book is to describe in sufficient detail the mathematical models of hysteresis nonlinearities with "nonlocal memories." The distinct feature of these nonlinearities is that their future states depend on past histories of input variations. It turns out that memories of hysteresis nonlinearities are quite selective. Indeed, experiments show that only some past input extrema (not the entire input variations) leave their marks upon future states of hysteresis nonlinearities. Thus, special mathematical tools are needed to describe nonlocal selective memories of hysteresis nonlinearities. The origin of such tools can be traced back to the landmark paper of Preisach.

The book is primarily concerned with Preisach-type models of hysteresis. All these models have a common generic feature; they are constructed as superpositions of simplest hysteresis nonlinearities—rectangular loops. The discussion is by and large centered around the following topics: various generalizations and extensions of the classical Preisach model (with special emphasis on vector generalizations), finding of necessary and sufficient conditions for the representation of actual hysteresis nonlinearities by various Preisach-type models, solution of identification problems for these models, and numerical implementation and experimental testing of Preisach-type models. Although the study of Preisach-type models constitutes the main subject of the book, some effort is also made to establish some interesting connections between these models and such topics as the critical state model for superconducting hysteresis, the classical Stoner-Wohlfarth model for vector magnetic hysteresis, thermal activation-type models for viscosity, magnetostrictive hysteresis and neural networks.

To make the book accessible to a broader audience, the exposition of the material is purposely made somewhat wordy with the hope that this will contribute to the clarity. This is especially true for the first chapter, while the discussion in the second and third chapters is more concise. The mathematical tools used in the book are quite simple. The exception is the discussion of the identification problems for the vector Preisach models in the third chapter, where some machinery of integral equations and the theory of irreducible representations of the group of rotations is occasionally used. No attempt is made to refer to all relevant publications. For this reason, the reference list is not exhaustive but rather suggestive; the presentation of the material in the book is largely based on my publications which are given at the end of the reference list.

I first heard about the Preisach model during my conversation with Professor K.M. Polivanov. This was about seventeen years ago, and at that time I lived in Russia. Shortly thereafter, my interest in the Preisach model was strongly enhanced as a result of brief discussions I had with Professor M.A. Krasnoselskii and his colleague A. Pokrovskii. When I came to the United States, my work on hysteresis modelling was strongly encouraged and supported by Dr. O. Manley from the U.S. Department of Energy. My research on the Preisach models has also strongly benefitted from many penetrating discussions I have had with Professor D. Fredkin (University of California, San Diego) over the past five years. I have also been fortunate to have such wonderful graduate students as T. Doong, G. Friedman, C. Korman, and A. Adly, who have assisted me at different times in my work on hysteresis. I am very thankful to Mrs. P. Keehn who patiently and diligently typed several versions of the manuscript. Finally, I gratefully acknowledge the financial support for my research on hysteresis from the U.S. Department of Energy, Engineering Research Program.

Contents

Preface ... vii
Introduction ... xi

Chapter I. The Classical Preisach Model of Hysteresis ... 1

1 Definition of the Classical Preisach Model ... 1
2 Geometric Interpretation and Main Properties of the Preisach Model ... 7
3 Experimental Determination of $\mu(\alpha, \beta)$. Representation Theorem ... 18
4 Numerical Implementation of the Preisach Model ... 32
5 The Preisach Model and Hysteretic Energy Losses ... 44
6 Preisach Model with Stochastic Input as a Model for Viscosity ... 55

Chapter II. Generalized Scalar Preisach Models of Hysteresis ... 64

1 Moving Preisach Model of Hysteresis ... 64
2 Nonlinear Preisach Model of Hysteresis ... 70
3 Restricted Preisach Models of Hysteresis ... 87
4 Dynamic Preisach Models of Hysteresis ... 99
5 Preisach Model of Hysteresis with Accommodation ... 108
6 Superconducting Hysteresis and Preisach-Type Models ... 114
7 Magnetostrictive Hysteresis and Preisach Models with Two Inputs 122
8 Experimental Testing of Preisach-Type Models of Hysteresis ... 129

Chapter III. Vector Presiach Models of Hysteresis ... 141

1 Classical Stoner-Wohlfarth Model of Vector Hysteresis ... 141
2 Definition of Vector Preisach Models of Hysteresis and Their Numerical Implementation ... 149
3 Some Basic Properties of Vector Preisach Hysteresis Models ... 157
4 Identification Problem for Isotropic Vector Preisach Models ... 172
5 Identification Problem for Anisotropic Vector Preisach Models ... 179
6 Dynamic Vector Preisach Models of Hysteresis ... 188
7 Generalized Vector Preisach Models of Hysteresis ... 194

References ... 202
Index ... 205

Introduction

The topic being discussed in this book is mathematical models of hysteresis. Special emphasis is placed on the mathematical exposition of these models which makes them quite general and applicable to the description of hysteresis of different physical nature. There are, however, two additional reasons for this emphasis. As was pointed out by A. Einstein [1], "mathematics enjoys special esteem, above all other sciences [because] its laws are absolutely certain and indisputable . . . ." Mathematics has achieved and maintained this exceptional position because its results are derived from a few (more or less self-evident) axioms by a chain of flawless reasonings. Since it is based on impeccable logic, mathematics can provide some level of security (and clarity) for natural sciences which is not attainable otherwise. For this reason, the rigorous mathematical representation of natural sciences is highly dcsirable and should be attempted whenever it is possible. In addition, mathematics more and more often serves as a vehicle of communication between scientists and engineers of different specializations. As a result, if some area of science is represented in a rigorous mathematical form, its accessibility is strongly enhanced. With these thoughts in mind, it is hoped that the mathematical exposition of hysteresis models undertaken here will bring much needed clarity into this area and will make it appealing to the broader audience of inquiring researchers.

This monograph has been written by an engineer for engineers. For this reason, mathematics is largely used here as a tool for achieving clarity. As a result, many mathematical subtleties of hysteresis modelling are omitted. These subtleties are by and large related to the fact that hysteresis operators are naturally defined on sets of piecewise monotonic functions which do not form complete function spaces. This leads to the problem of continuous extension of hysteresis operators from the above sets to some complete function spaces. The reader interested in this type of mathematical problems is referred to the study by the Russian mathematicians M. Krosnoselskii and A. Pokrovskii [2].

The phenomenon of hysteresis has been with us for ages and has been attracting the attention of many investigators for a long time. The reason is that hysteresis is ubiquitous. It is encountered in many different areas of science. Examples include magnetic hysteresis, ferroelectric hysteresis, mechanical hysteresis, superconducting hysteresis, adsorption hysteresis, optical hysteresis, electron beam hysteresis, etc. However, the very meaning of hysteresis varies from one area to another, from paper to paper, and from author to author. As a result, a stringent mathematical definition of hysteresis is needed in order to avoid confusion and ambiguity. Such a definition will serve a twofold purpose: first, it will be a substitute for vague notions, and, second, it will pave the road for more or less rigorous proofs.

We begin with the definition of scalar hysteresis and, for the sake of generality, we adopt the language of control theory. Consider a transducer (see Fig. 1) which can be characterized by an input $u(t)$ and an output $f(t)$. This transducer is called a hysteresis transducer (HT) if its input-output relationship is a multibranch nonlinearity for which branch-to-branch transitions occur after input extrema. This multibranch nonlinearity is shown in Fig. 2. For the most part, the case of static hysteresis nonlinearity will be discussed. The term "static" means that branches of such hysteresis nonlinearities are determined only by the past extremum values of input, while the speed (or particular manner) of input variations between extremum points has no influence on branching. This statement is illustrated by Figs. 3a, 3b, and 3c. Figures 3a and 3b show two different inputs $u_1(t)$ and $u_2(t)$ which successively assume the same extremum values but vary differently between these values. Then, for a static HT, these two inputs will result in the same $f - u$

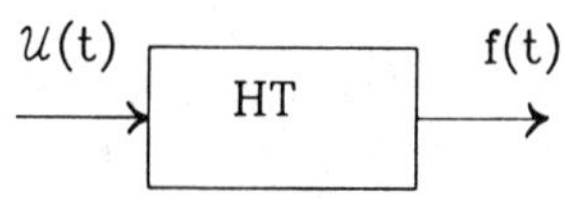

Fig. 1

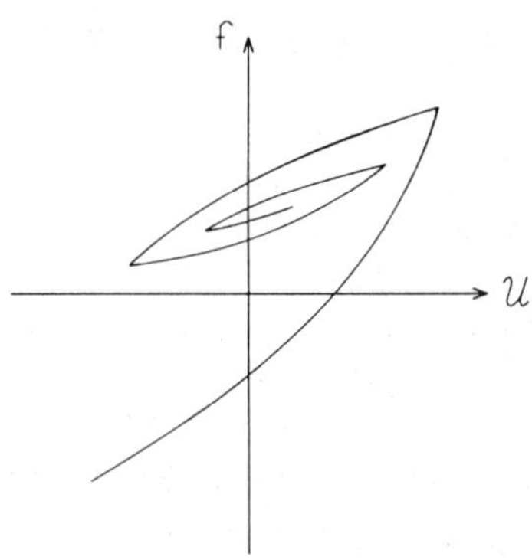

Fig. 2

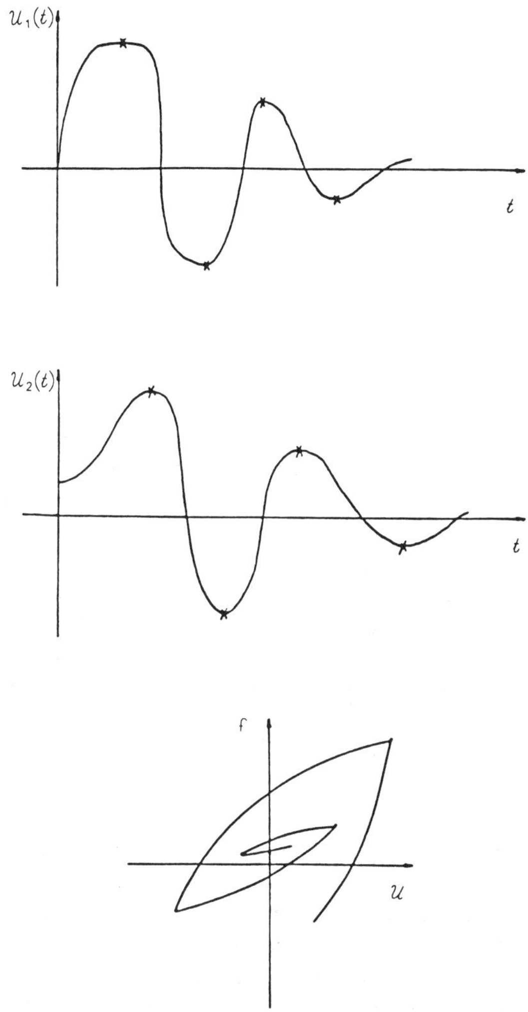

Fig. 3

diagram (see Fig. 3c), provided that the initial state of the transducer is the same for both inputs.

The given definition of static scalar hysteresis is consistent with existing experimental facts. Indeed, it is known in the area of magnetic hysteresis that a shape of major (or minor) loop (see Fig. 4) can be specified without referring to how fast magnetic field H varies between two extremum values, $+H_m$ and $-H_m$. This indicates that time effects are negligible and the given definition of a static hysteresis transducer is an adequate one. It is worthwhile to keep in mind that, for very fast input variations, time effects become important and

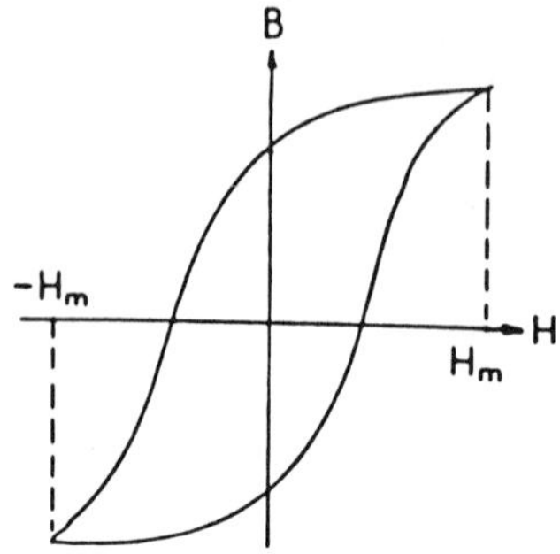

Fig. 4

the given definition of static hysteresis fails. In other words, this definition (as any other definition) has its limits of applicability to real life problems.

In the existing literature, the hysteresis phenomenon is by and large linked with the formation of hysteresis loops (looping). This may be misleading and create the impression that looping is the essence of hysteresis. In this respect, the given definition of hysteresis emphasizes the fact that branching constitutes the essence of hysteresis, while looping is a particular case of branching. Indeed, looping occurs when the input varies back and forth between two consecutive extremum values, while branching takes place for arbitrary input variations.

From the given definition, it can also be concluded that scalar hysteresis can be interpreted as a nonlinearity with a memory which reveals itself through branching.

In the given definition of hysteresis, the physical meanings of the input $u(t)$ and the output $f(t)$ were left unspecified. This was done deliberately, for the sake of mathematical generality. However, it is not difficult to specify the meanings of $u(t)$ and $f(t)$ in particular applications. For instance, in magnetism $u(t)$ is the magnetic field and $f(t)$ is the magnetization, in mechanics $u(t)$ is the force and $f(t)$ is the displacement (length), in adsorption $u(t)$ is the gas pressure and $f(t)$ is the amount of material adsorbed. The notion of a hysteresis transducer may have different interpretations as well. For instance, in magnetism the HT can be construed as an infinitesimally small volume of magnetic material, and the corresponding input-output hysteresis nonlinearity can be interpreted as a constitutive equation for this material.

All static hysteresis nonlinearities fall into two general classifications: (a) hysteresis nonlinearties with local memories and (b) hysteresis nonlinearities with nonlocal memories. The hysteresis nonlinearities with local memories are characterized by the following property. The value of output $f(t_0)$ at some instant of time t_0 and the values of input $u(t)$ at all subsequent instants of time $t \geq t_0$ uniquely predetermine the value of output $f(t)$ for all $t > t_0$. In other words, for hysteresis transducers with local memories the past exerts its influence upon the future through the current value of output. This is not the

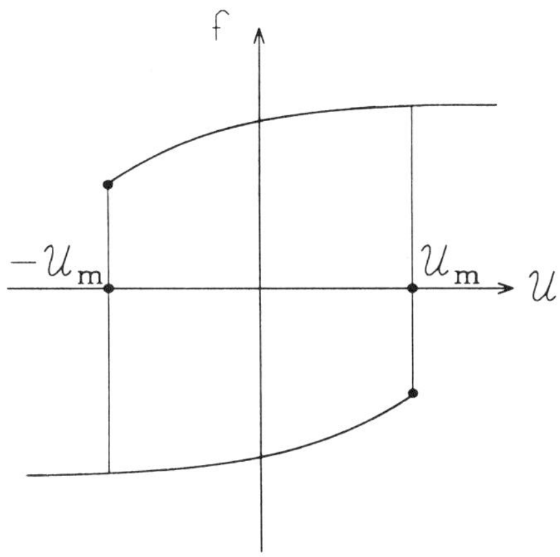

Fig. 5

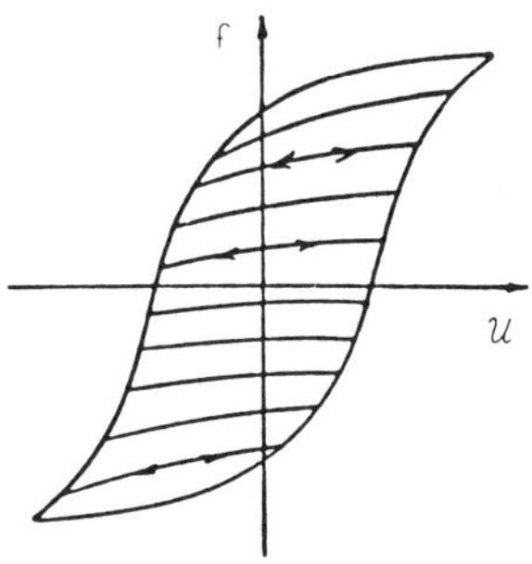

Fig. 6

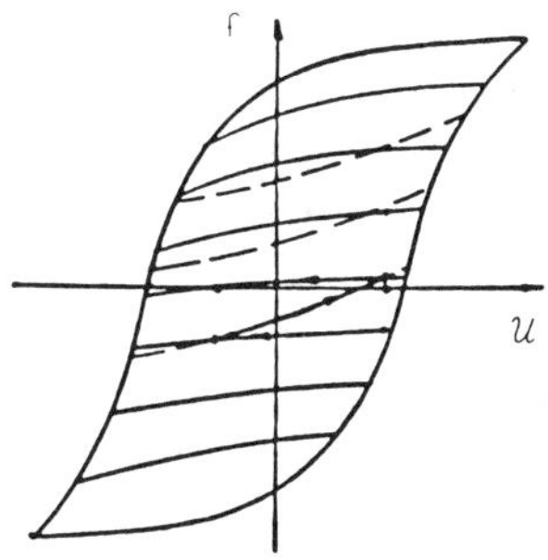

Fig. 7

case for hysteresis transducers with nonlocal memories. For such transducers, future values of output $f(t)$, $(t \geq t_0)$ depend not only on the current value of output $f(t_0)$ but on past extremum values of input as well.

The typical examples of hysteresis nonlinearities with local memories are shown in Figs. 5, 6, and 7. Figure 5 shows the simplest hysteresis nonlinearity with local memory. It is specified by a major loop which is formed by ascending and descending branches. These branches are only partially reversible (their vertical sections are not reversible). This type of hysteresis nonlinearity is characteristic, for instance, for single Stoner-Wohlfarth magnetic particles [3]. For this type of hysteresis, branching occurs if only extremum values of input exceed $+u_m$ or $-u_m$.

A more complicated type of hysteresis nonlinearities with local memories is illustrated by Fig. 6. Here, there is a set of inner curves within the major loop and only one curve passes through each point in the $f - u$ diagram. These curves are fully reversible and can be traversed in both directions, for a monotonically increasing and decreasing input $u(t)$. For this type of hysteresis, branching may occur only when ascending or descending branches of major loops are reached.

A hysteresis nonlinearity with local memory which has two sets of inner curves (the ascending and descending curves) is shown in Fig. 7. This type of hysteresis was probably first described by Madelung [4] at the beginning of the century, and afterwards it was independently invented by many authors time and time again (see, for instance, [5] and [6]). For this hysteresis nonlinearity, only one curve of each set passes through each point in the $f - u$ diagram. If the input $u(t)$ is increased, the ascending curve is followed, if it is decreased, the descending curve is traced. Thus, branching occurs for any input extremum. However, in general, minor loops are not formed; if $u(t)$ varies back and forth between the same two values, the output usually exhibits a continued upward drift (see Fig. 7).

It is clear from the above examples that all hysteresis nonlinearities with local memories have the following common feature: every reachable point in the $f - u$ diagram corresponds to a uniquely defined state. This state predetermines the behavior of HT in exactly one way for increasing $u(t)$ and exactly one way for decreasing $u(t)$. In other words, at any point in the $f - u$ diagram there are only one or two curves that may represent the future behavior of HT with local memory (see Fig. 8). This is not true for hysteresis transducers with nonlocal memories. In the latter case, at any reachable point in the $f - u$ diagram there is an *infinity* of curves that may represent the future behavior of the transducer (see Fig. 9). Each of these curves depends on a particular past history, namely, on a particular sequence of past extremum values of input. By analogy with the random process theory, hysteresis nonlinearities with local memories can be called markovian hysteresis nonlinearities, while hysteresis nonlinearities with nonlocal memories are nonmarkovian. It is clear that hysteresis nonlinearities with nonlocal memories are much more complicated than those with local memories.

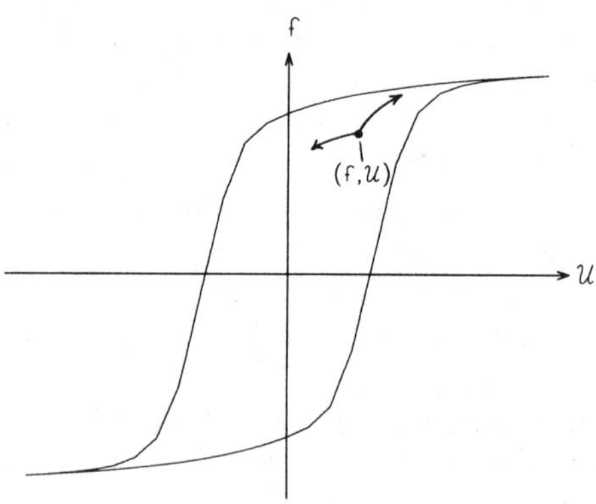

Fig. 8

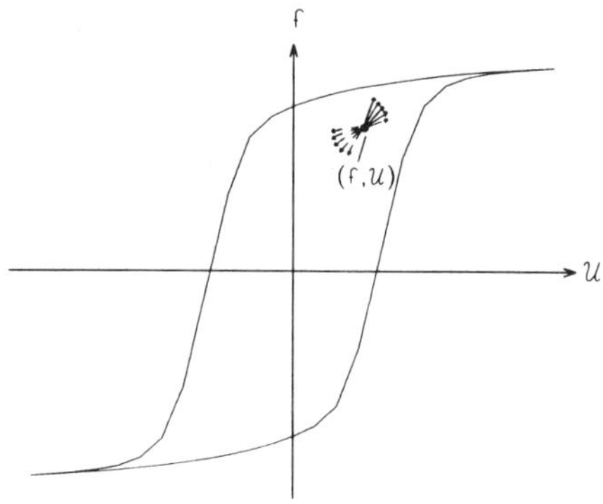

Fig. 9

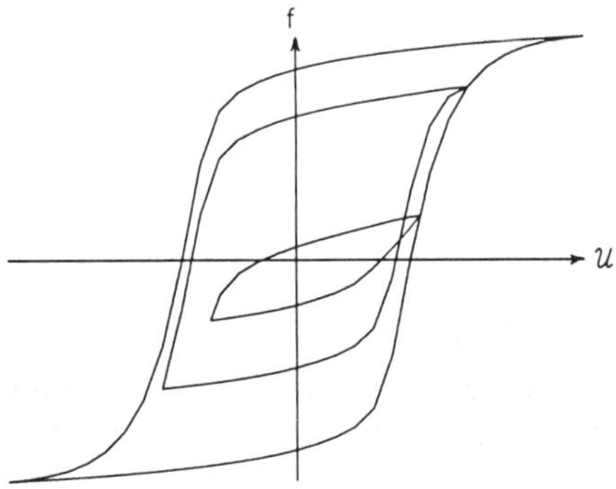

Fig. 10

Mathematical models of hysteresis nonlinearities with local memories have been extensively studied by using differential and algebraic equations. These models have achieved a high level of sophistication which is reflected, for instance, in publications [7–10]. However, the notion of hysteresis nonlinearities with local memories is not consistent with experimental facts. For instance, it is reported in [21] that crossing and partially coincident minor loops have been experimentally observed. These loops are schematically shown in Fig. 10 and 11, respectively. The existence of crossing minor loops attached to a major loop is more or less obvious, while the presence of partially coincident minor loops is a more subtle phenomenon. The existence of crossing and partially coincident minor loops clearly suggests that the states of the corresponding hysteresis transducers are not uniquely specified by their inputs and outputs. Thus, hysteresis of this transducer does not have a local memory.

This book is solely concerned with mathematical models of hysteresis with nonlocal memory. The question arises, why are these models needed? The answer is that the hysteresis transducer is usually a part of a system. As a

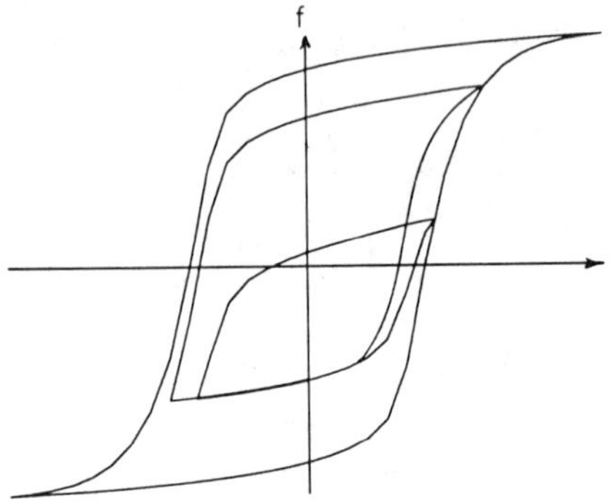

Fig. 11

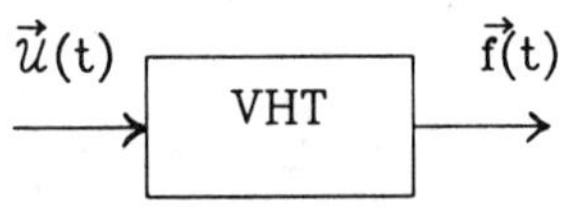

Fig. 12

result, its input is not known beforehand, but is determined by the interaction of the transducer with the rest of the system. Since the input of HT is not predictable a priori, it is impossible to specify ahead of time the branches of hysteresis nonlinearity which will be followed in a particular regime of the system. This is the main impediment as far as self-consistent mathematical descriptions of systems with hysteresis are concerned. To overcome the difficulty mentioned above, mathematical models of hysteresis are needed. These models represent new mathematical tools which themselves (due to their structure) will detect and accumulate input extrema and will choose appropriate branches of the hysteresis nonlinearity with respect to the accumulated histories. Coupled together with mathematical description of the rest of the system these models will constitute complete and self-consistent mathematical descriptions of systems with hysteresis. Without such models, the self-consistent mathematical descriptions of systems with hysteresis are virtually impossible.

We next turn to the discussion of vector hysteresis. This hysteresis can be characterized by a vector input $\vec{u}(t)$ and vector output $\vec{f}(t)$ (see Fig. 12). Two- and three-dimensional vector inputs and vector outputs are most relevant to practical applications. That is why only two- and three-dimensional vector hysteresis models are discussed in this book. However, the formal mathematical generalization of these models to n dimensions ($n > 3$) is straightforward. It is believed that such a generalization will be performed by the reader if needed.

The most immediate problem we face is how to define vector hysteresis in a mathematically rigorous as well as physically meaningful way. To do this, it is important to understand what constitutes the essential part of past input history in the case of vector hysteresis. In other words, we need to understand what part of the past input history affects the future values of output. To resolve this problem, it would be natural to turn to experimental facts. Unfortunately, the experimental data available in the area of vector hysteresis are quite scarce and inconclusive. The latter means that these data provide no clear indications as to how the problem posed above can be resolved. Nevertheless, it is possible to find some resolution of this problem by strictly logical reasoning. The starting point for our inference is the notion that scalar hysteresis is a particular case of vector hysteresis. This case is realized when the vector input $\vec{u}(t)$ is restricted to vary along only one direction. In fact, it can be successfully argued (at least in the area of magnetics) that there is no such a thing as scalar hysteresis. Whenever we talk about scalar hysteresis, we are actually dealing with some specific properties of vector hysteresis which have been observed for input variations along some fixed directions. Thus, it can be concluded that vector hysteresis is reduced to scalar hysteresis when the vector input is restricted to vary along an arbitrarily fixed direction. Since the above reduction occurs for *any* fixed direction and since past extremum values of input constitute the essential part of past history for scalar hysteresis, it is natural to conclude that in the case of vector hysteresis past extremum values of input projections along all possible directions may leave their mark upon the future. As a result, we arrive at the following definition of vector hysteresis. Vector hysteresis is a vector nonlinearity with the property that past extremum values of input projections along all possible directions may affect future values of output. Since no mention is made concerning the speed of input variations, it is apparent that the above definition implies static vector hysteresis. It is also clear that mathematical models of vector hysteresis are imperative for self-consistent descriptions of systems with vector hysteresis. These models should be able to detect and store past extrema of input projections along all possible directions and choose the appropriate value of vector output according to the accumulated history.

This book deals exclusively with the mathematical models of hysteresis which are purely phenomenological in nature. Essentially, these models represent the attempt to describe and generalize experimental facts. They provide no insights into specific physical causes of hysteresis. Nevertheless, they have been and may well continue to be powerful tools for device design. There are, however, fundamental models of hysteresis which attempt to explain experimental facts from first principles. For instance, in micromagnetics, these principles require that the equilibrium distribution of magnetization should correspond to an energy minimum. The minimized energy basically includes the exchange energy, the anisotropy energy, the energy of interaction with an applied field, the magnetostatic self-energy, and possibly some other terms. It turns out that there are many (at least two) different local minima of the total energy for a given applied field. Since only one of these energy minima corresponds to the thermodynamic equilibrium state, the others must be metastable. They may persist for very long times. These pcrsisting metastable states are responsible for the origin of hysteresis.

Although the above micromagnetic approach is fundamental in nature, its implementation encounters some intrinsic difficulties.

First, in order to carry out this approach, the detailed information of microscopic material structure is needed. Only on the basis of this information, can the above-mentioned terms of minimized energy be specified. However, the detailed knowledge of material microstructure is often not available.

Second, the micromagnetic approach leads to nonlinear differential (or integrodifferential) equations which are quite complicated to solve even using sophisticated numerical techniques. In part, this is because the solution of these equations may exhibit highly irregular behavior. Indeed, domains and their walls should emerge from the micromagnetic approach. The domain walls are small regions where the direction of magnetization changes quite rapidly, from some particular direction in one domain to a different direction in an adjacent domain. In a way, these domain walls can be mathematically construed as interior layers. This suggests that micromagnetic problems may well belong to the class of singularly perturbed problems. (This fact has not been appreciated enough in the existing literature.) To resolve the fast variations of magnetization over the domain walls, very fine meshes are needed. But, the domain walls usually move when the applied field is changed. Thus, it is not clear a priori where the fine meshes should be located. This may seriously complicate the numerical analysis.

Finally, the detailed domain structure which can be produced by the micromagnetic approach may be irrelevant to some practical problems. This is the case, for instance, in the design of devices for which the average value of magnetization over regions with dimensions much larger than domain dimensions is of interest.

Summarizing the above discussion, it can be concluded that the phenomenological approach is more directly connected with macroscopic experimental data. For this reason, it is of a great value to the device design. The fundamental micromagnetic approach, on the other hand, is intimately related to material structure and, therefore, it can be useful in the design of new materials.

CHAPTER 1

The Classical Preisach Model of Hysteresis

1 Definition of the Classical Preisach Model

This model has a long and instructive history which can be best characterized by the following eloquent statement of J. Larmor made in the preface to the book [11]:

... scientific progress, considered historically, is not a strictly logical process, and does not proceed by syllogisms. New ideas emerge dimly into intuition, come into consciousness from nobody knows where, and become the material on which the mind operates, forging them gradually into consistent doctrine, which can be welded on to existing domains of knowledge.

This is exactly what has happened with the Preisach model. The origin of this model can be traced back to the landmark paper of F. Preisach [12] published in 1935. Preisach's approach was purely intuitive. It was based on some plausible hypotheses concerning the physical mechanisms of magnetization. For this reason, the Preisach model was first regarded as a physical model of hysteresis. It has remained known primarily in the area of magnetics where this model has been the focus of considerable research for many years. This has resulted in the further development of the Preisach model, and many improvements and valuable facts have been accumulated in this area. These developments and results are recorded, for instance in [13–22]; it is important to note that the given list of references is not complete by any standards.

Somewhat in parallel with the mentioned developments in magnetics, the Preisach model was independently invented and then extensively studied and tested for adsorption hysteresis by D.H. Everett and his collaborators [23–26]. This clearly indicated that the applications of Preisach's model were not limited only to the area of magnetics.

The next decisive step in the direction of better understanding of the model was made in the 1970s by the Russian mathematician M. Krasnoselskii who realized that the Preisach model contained a new general mathematical idea. Krasnoselskii separated this model from its physical meaning and represented it in a purely mathematical form which is similar to a spectral decompo-

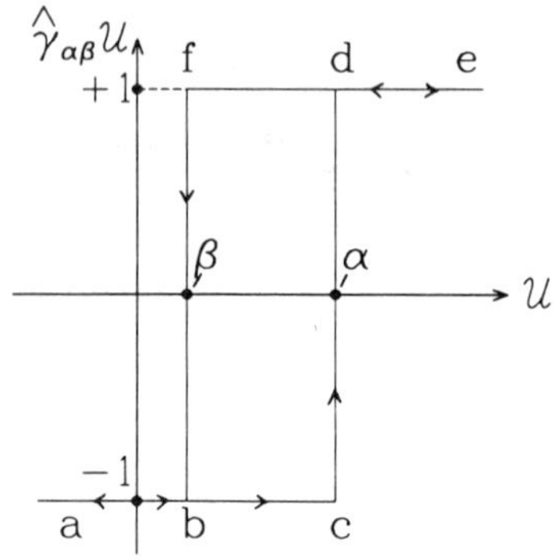

Fig. 1.1

sition of operators [2]. As a result, a new mathematical tool has evolved which can now be used for the mathematical description of hysteresis of different physical nature. At the same time, the Krasnoselskii approach has strongly revealed the phenomemological nature of the Preisach model. This has put this model into a new perspective and has led to the clarification of many controversies which have long surrounded the Preisach model.

We next proceed to the purely mathematical description of the Preisach model. Consider an infinite set of simplest hysteresis operators $\hat{\gamma}_{\alpha\beta}$. Each of these operators can be represented by a rectangular loop on the input-output diagram (see Fig. 1.1). Numbers α and β correspond to "up" and "down" switching values of input, respectively. It will be assumed in the sequel that $\alpha \geq \beta$ which is quite natural from the physical point of view. Outputs of the above elementary hysteresis operators may assume only two values $+1$ and -1. In other words, these operators can be interpreted as two-position relays with "up" and "down" positions corresponding to $\hat{\gamma}_{\alpha\beta}u(t) = +1$ and $\hat{\gamma}_{\alpha\beta}u(t) = -1$, respectively. As the input $u(t)$ is monotonically increased, the ascending branch *abcde* is followed. When the input is monotonically decreased, the descending branch *edfba* is traced. It is clear that the operators $\hat{\gamma}_{\alpha\beta}$ represent hysteresis nonlinearities with local memories (see the Introduction). Along with the set of operators $\hat{\gamma}_{\alpha\beta}$ consider an arbitrary weight function $\mu(\alpha, \beta)$ which is often referred to as the Preisach function. Then, the Preisach model can be written as

$$f(t) = \hat{\Gamma}u(t) = \iint_{\alpha \geq \beta} \mu(\alpha, \beta)\hat{\gamma}_{\alpha\beta}u(t)\,d\alpha\,d\beta. \tag{1.1}$$

Here $\hat{\Gamma}$ is used for the concise notation of the Preisach hysteresis operator which is defined by the integral in (1.1).

It is apparent that the model (1.1) can be interpreted as a continuous analog of a system of parallelly connected two-position relays. This interpretation is illustrated by the block diagram shown in Fig. 1.2. According to this diagram, the same input $u(t)$ is applied to each of the two-position relays. Their individual outputs are multiplied by $\mu(\alpha, \beta)$ and then integrated over all appro-

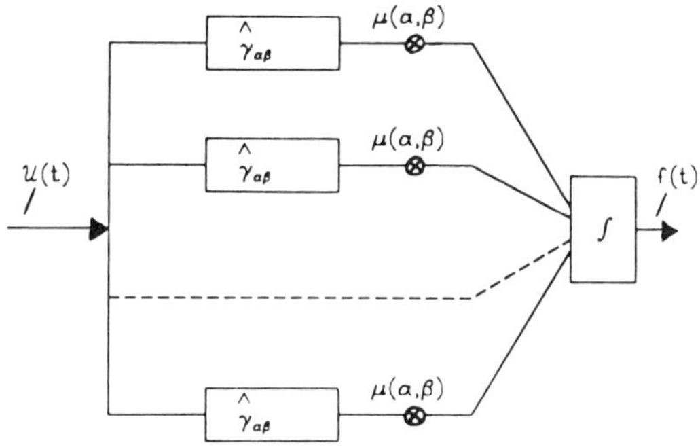

Fig. 1.2

priate values of α and β. As a result, the output $f(t)$ is obtained. Discrete approximation to the above block diagram can be used as device realizations of the Preisach model (1.1).

It is clear from the above discussion that the Preisach model is constructed as a superposition of simplest hysteresis operators $\hat{\gamma}_{\alpha\beta}$. These operators can be construed as main building blocks for the model (1.1). The idea that a complicated operator can be represented as a superposition of simplest operators is not entirely new and was exploited before in mathematics, particularly in the functional analysis [27]. For instance, according to the spectral decomposition theory for self-adjoint operators, any self-adjoint operator can be represented as a superposition of projection operators which are, in a way, the simplest self-adjoint operators. The above analogy shows that the Preisach model (1.1) can be interpreted from the mathematical point of view as a spectral decomposition of complicated hysteresis operator $\hat{\Gamma}$ into the simplest hysteresis operators $\hat{\gamma}_{\alpha\beta}$. This interpretation has been emphasized by Krasnoselskii [2].

Although the Preisach hysteresis nonlinearity (1.1) is constructed as a superposition of elementary hysteresis nonlinearities $\hat{\gamma}_{\alpha\beta}$ with local memories, it usually has a nonlocal memory. (This fact will be proven in the next section). It is remarkable that a new qualitative property of nonlocal memory emerges as a collective property of a system having huge (infinite) number of simple and qualitatively similar components.

Having defined the Preisach model, it is appropriate to describe the directions along which the further discussion will proceed. The subsequent discussion in this chapter will be centered around the following six issues.

1. How does the Preisach model work? In other words, how does this model detect local input extrema, accumulate them, and choose the appropriate branches of hysteresis nonlinearity with respect to the accumulated histories?

The answer to this question will reveal the mechanism of memory formation in the Preisach model. It will require the development of a special diagram technique which will constitute the mathematical foundation for the investigation of the Preisach model.

2. What experimental data are needed for the determination of the Preisach function $\mu(\alpha, \beta)$ for a given hysteresis transducer?

This is a so-called identification problem. The solution to this problem is very important as far as the practical applications of the Preisach model are concerned.

3. What are the necessary and sufficient conditions for the representation of actual hysteresis nonlinearities by the Preisach model?

The significance of this problem is that its solution will clearly establish the limits of applicability of the Preisach model.

4. How can the Preisach model be implemented numerically?

This is an important question because it seems at first that the numerical evaluation of two-dimensional integrals is required for the numerical implementation of the Preisach model. However, it will be shown in the sequel that the evaluation of the above integrals can be completely avoided.

5. How can the Preisach model be useful for the computation of hysteretic energy dissipation?

It is well knwon that the hysteretic energy dissipation can be easily evaluated for the case of periodic (cyclic) input variations. However, the problem of computing hysteretic energy losses for arbitrary (not necessarily periodic) input variations has remained unsolved. It will be shown in this chapter (see Section 5) that the Preisach model can bring about the solution to this problem. This solution can be useful for the computation of internal entropy production and, in this way, may facilitate the development of the irreversible thermodynamics of hysteretic media.

6. Can the Preisach model be useful for the description of viscosity phenomena in hysteretic systems?

It is well known that the physical origin of hysteresis is the multiplicity of metastable states. At equilibrium, large deviations of random thermal perturbations may cause hysteretic systems to move from higher to lower energy metastable states resulting in slow changes of output variables. This phenomenon is known in the literature as "viscosity." It will be shown in this chapter (see Section 6) that the Preisach model with stochastic input can be used for the description of the viscosity phenomenon. This approach will be compared with traditional thermal activation-type models.

There is another problem which has not been mentioned above but has been extensively studied by Krasnoselskii and his colleagues [2], [28]. This problem is related to the fact that hysteresis nonlinearities are naturally defined on the set of continuous and piecewise monotonic inputs. However, the above set of functions does not form a complete function space. This constitutes the main difficulty as far as the rigorous mathematical treatment of differential (or integral) equations with hysteresis nonlinearities is concerned. Thus, the problem of continuous extension of hysteresis operators from the above set of piecewise monotonic inputs to some complete function spaces presents itself. The essence of the problem is in finding (or describing) such complete function spaces. The solution to this problem is important because these function spaces form the natural "environment" for the rigorous mathematical study of equations with hysteresis nonlinearities. Nevertheless, this problem is more or less of purely mathematical nature. It is not directly related to practical applications, and for this reason will not be discussed in the book. The reader interested in the discussion of this problem is referred to Krasnoselskii and Pokrovskii [2] as well as some other recent publications [29–34].

It is apparent from the above discussion that the Preisach model has been defined without any reference to a particular physical origin of hysteresis. This clearly reveals the phenomenological nature of the model and its mathematical generality. To further emphasize this new approach to the Preisach model, it is instructive to compare the definition of the model given above with the definition regularly used in magnetics.

In magnetics, separate magnetic "particles" (or "domains") are introduced. These particles have rectangular hysteresis loops and they play the same role as elementary hysteresis operators $\hat{\gamma}_{\alpha\beta}$. A loop typical for such particles is shown in Fig. 1.3. Here H_u and H_d are "up" and "down" switching magnetic fields, respectively, and m_s is the magnetic moment (magnetization) which is identical for all particles. The notation $\hat{m}(H_u, H_d)$ is used for the particle having the hysteresis loops shown in Fig. 1.3. The magnetic material is considered to be composed of many such particles. It is also assumed that the different particles have some distribution of reversal field H_u and H_d which can be characterized by the distribution function $\phi(H_u, H_d)$. This function plays the same role as $\mu(\alpha, \beta)$ in (1.1). It is typical to speak about the statistical nature of the distribution function $\phi(H_u, H_d)$, although the author is not aware of any rigorous justification or experimental evidence for this statistical interpretation. By using the magnetic particles and their distribution function, the

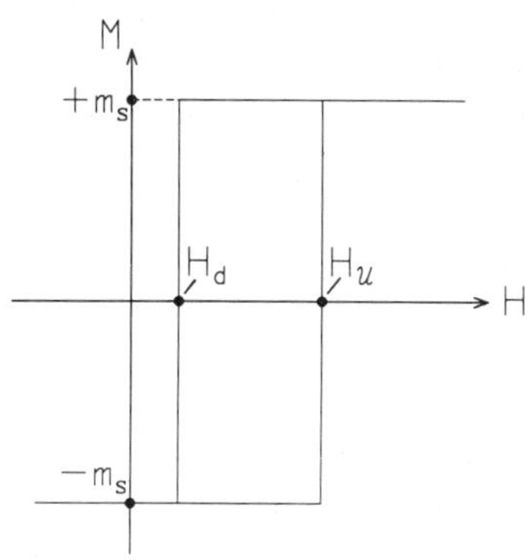

Fig. 1.3

Preisach model is usually defined in magnetics as

$$M(t) = \iint_{H_u \geq H_d} \phi(H_u, H_d)\hat{m}(H_u, H_d)H(t)dH_u dH_d, \tag{1.2}$$

where M is the magnetization,

$$\hat{m}(H_u, H_d)H(t) = +m_s \tag{1.3}$$

if the particle is swiched up, and

$$\hat{m}(H_u, H_d)H(t) = -m_s \tag{1.4}$$

if the particle is switched down.

The given definition is framed in terms of magnetics and can naturally be called the "magnetic" definition. It is obvious that the "magnetic" definition of the Preisach model is mathematically equivalent to the previous definition (1.1).

The "magnetic" definition suggests some relation of the Preisach model to certain physical realities such as magnetic particles, their distribution, etc. For this reason, the Preisach model has long been regarded as the physical model. However, there are some intrinsic difficulties with respect to this interpretation. Indeed, if the magnetic particles are to have some features of reality, then their geometric shapes and mutual spatial locations should be important and affect the value of magnetization. But, in the "magnetic" definition of the Preisach model there is no reference whatsoever to the spatial locations of the particles or to their shapes. In the mathematical definition (1.1), this difficulty does not appear because there is no need to relate the operators, $\hat{\gamma}_{\alpha\beta}$, to some real physical objects.

Another difficulty comes from the fact that many (almost all) particles in the Preisach model have asymmetrical rectangular loops ($H_u \neq -H_d$). Such loops are not physical for single particles. The use of asymmetrical loops is commonly justified on the ground that asymmetry is caused by the interaction between the particles, and the amount of asymmetry is a measure of the interaction. More precisely, this means that each particle "feels" not only the applied (external) magnetic field $H(t)$, but also the interaction magnetic field which is due to the magnetization of adjacent particles. The interaction magnetic field may, of course, vary from one particle to another. These interaction magnetic fields result in shifted (asymmetrical) hysteresis loops when these loops represent the dependence of magnetization on the applied field alone.

Although the given explanation sounds plausible, it is not completely satisfactory. This explanation does not account for the fact that the interaction fields depends on particular states of magnetization, shapes and location of the particles. This dependence is not represented in the Preisach model at all. In the case of purely mathematical definition (1.1) of the Preisach model, the above diffiuclty does not appear. This is because there is a straightforward phenomenological explanation for the need of operators $\hat{\gamma}_{\alpha\beta}$ with asymmetri-

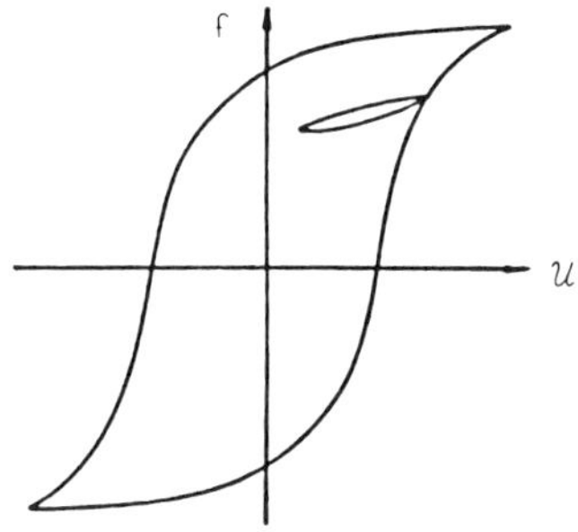

Fig. 1.4

cal loops. These operators are needed in order to describe asymmetrical minor loops exhibited by actual hysteresis nonlinearities (see Fig. 1.4). Indeed, if only operators $\hat{\gamma}_{\alpha,-\alpha}$ with symmetrical loops were used, then the model (1.1) being constructed as a superposition of such $\hat{\gamma}$-operators would also describe only symmetrical minor loops. In other words, $\hat{\gamma}$-operators with symmetrical loops do not form a complete set of operators in a sense that hysteresis nonlinearities with asymmetrical minor loops can not be represented as superpositions of $\hat{\gamma}$-operators from this set.

From the discussion presented above the following conclusion can be drawn. The "magnetic" definition of the Preisach model is possible and historically was first developed. However, this definition obscures the model, conceals its mathematical and phenomenological nature, and narrows the area of applicability of this model to the field of magnetics. The definition (1.1), on the other hand, interprets the Preisach model as a new mathematical tool whose importance may well extend beyond the area of magnetics. For this reason, the purely mathematical definition (1.1) seems to be more attractive and will be used throughout the book.

2 Geometric Interpretation and Main Properties of the Preisach Model

The mathematical investigation of the Preisach model is considerably facilitated by its geometric interpretation. This interpretation is based on the following simple fact. There is a one-to-one correspondence between operators $\hat{\gamma}_{\alpha\beta}$ and points (α, β) of the half-plane $\alpha \geq \beta$ (see Fig. 1.5). In other words, each point of the half-plane $\alpha \geq \beta$ can be identified with only one particular $\hat{\gamma}$-operator whose "up" and "down" switching values are respectively equal to α and β coordinates of the point. It is clear that this identification is possible because both $\hat{\gamma}$-operators and the points of the half-plane $\alpha \geq \beta$ are uniquely defined by pairs of numbers, α and β.

Consider a right triangle T (see Fig. 1.5). Its hypotenuse is a part of the line $\alpha = \beta$, while the vertex of its right angle has the coordinates α_0 and β_0 with $\beta_0 = -\alpha_0$. In the sequel, this triangle will be called the limiting triangle and

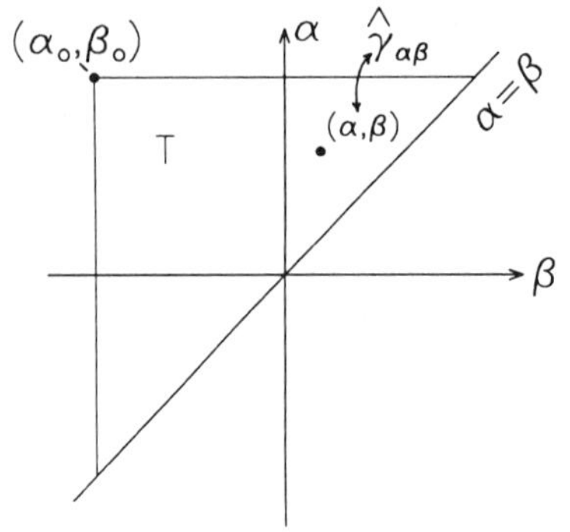

Fig. 1.5

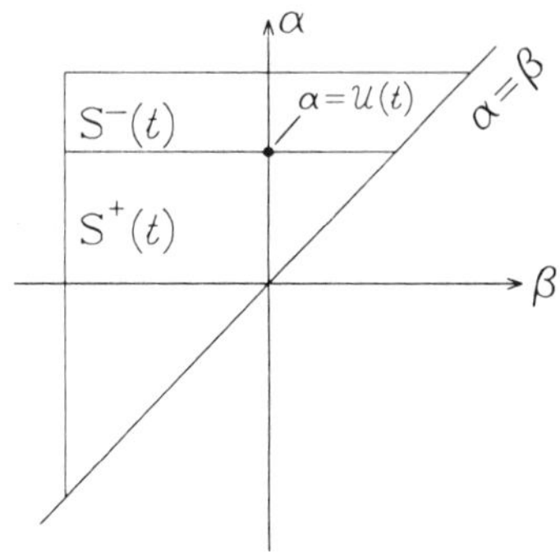

Fig. 1.6

the case when $\mu(\alpha, \beta)$ is a finite function with a support within T will be discussed. In other words, it will be assumed that the function $\mu(\alpha, \beta)$ is equal to zero outside the triangle T. This case covers the important class of hysteresis nonlinearities with closed major loops. At the same time, the above case will not essentially limit the generality of our future discussions.

To start the discussion, we first assume that the input $u(t)$ at some instant of time t_0 has the value which is less than β_0. Then, the outputs of all $\hat{\gamma}$-operators which correspond to the points of the triangle T are equal to -1. In other words, all $\hat{\gamma}$-operators are in the "down" position. This corresponds to the state of "negative saturation" of the hysteresis nonlinearity represented by the model (1.1).

Now, we assume that the input is monotonically increased until it reaches at time t_1 some maximum value u_1. As the input is being increased, all $\hat{\gamma}$-operators with "up" switching values α less than the current input value $u(t)$ are being turned into the "up" position. This means that their outputs become equal to $+1$. Geometrically, it leads to the subdivision of the triangle T into two sets: $S^+(t)$ consisting of points (α, β) for which the corresponding $\hat{\gamma}$-operators are in the "up" position, and $S^-(t)$ consisting of points (α, β) such that the corresponding $\hat{\gamma}$-operators are still in the "down" position. This subdivision is made by the line $\alpha = u(t)$ (see Fig. 1.6), which moves upwards as the input is being increased. This upward motion is terminated when the input reaches the maximum value u_1. The subdivision of the triangle T into $S^+(t)$ and $S^-(t)$ for this particular instant of time is shown in Fig. 1.7.

Next, we assume that the input is monotonically decreased until it reaches at time t_2 some minimum value u_2. As the input is being decreased, all $\hat{\gamma}$-operators with "down" switching values β above the current input value $u(t)$ are being turned back into the "down" position. This changes the previous subdivision of T into positive and negative sets. Indeed, the interface $L(t)$ between $S^+(t)$ and $S^-(t)$ now has two links, the horizontal and vertical. The vertical link moves from right to left and its motion is specified by the equation $\beta = u(t)$. This is illustrated by Fig. 1.8. The above motion of the vertical link

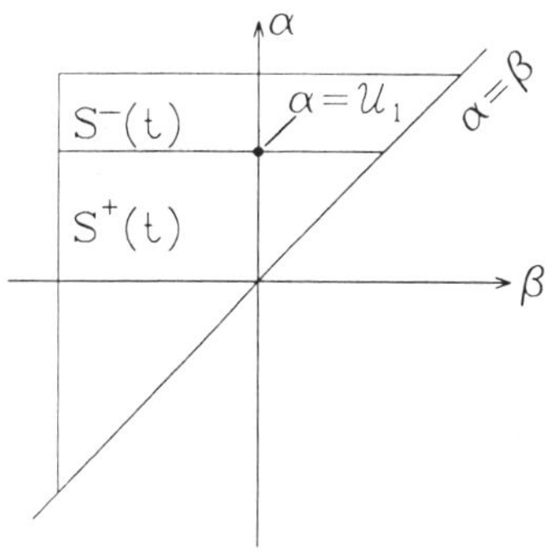

Fig. 1.7

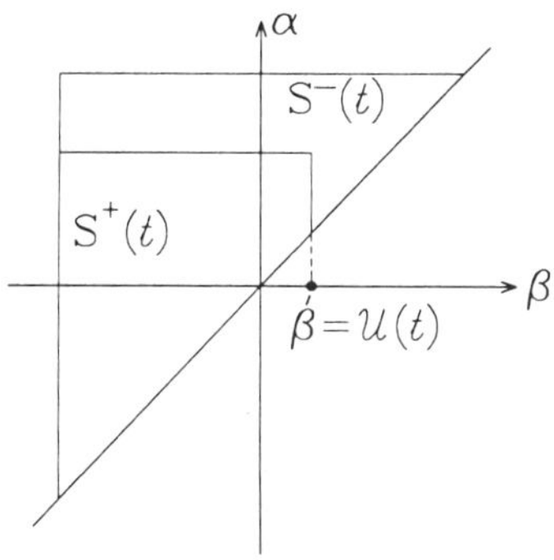

Fig. 1.8

is terminated when the input reaches its minimum value u_2. The subdivision of the triangle T for this particular instant of time is shown in Fig. 1.9. The vertex of the interface $L(t)$ at the above instant of time has the coordiates $\alpha = u_1$ and $\beta = u_2$.

Now, we assume that the input is increased again until it reaches at time t_3 some maximum value u_3 which is less than u_1. Geometrically, this increase results in the formation of a new horizontal link of $L(t)$ which moves up. This upward motion is terminated when the maximum u_3 is reached. This is shown in Fig. 1.10.

Next, we assume that the input is decreased again until it reaches at time t_4 some minimum value u_4 which is above u_2. Geometrically, this input variation results in the formation of a new vertical link which moves from right to left. This motion is terminated as the input reaches its minimum value u_4. As a result, a new vertex of $L(t)$ is formed which has the coordinates $\alpha = u_3$ and $\beta = u_4$. This is illustrated by Fig. 1.11.

By generalizing the previous analysis, the following conclusion can be reached. At anly instant of time, the triangle T is subdivided into two sets: $S^+(t)$ consisting of points (α, β) for which the corresponding $\hat{\gamma}$-operators are

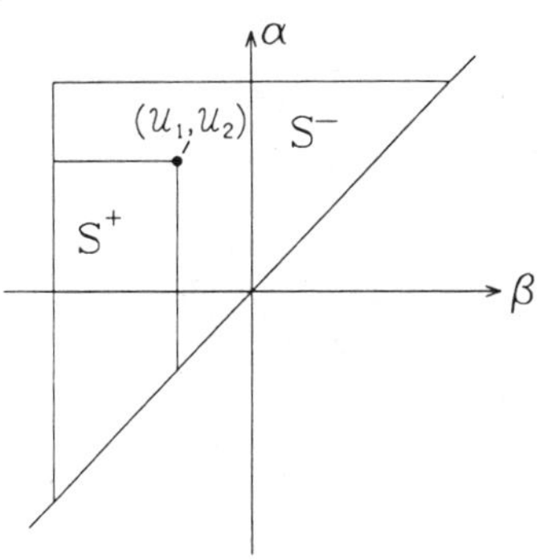

Fig. 1.9

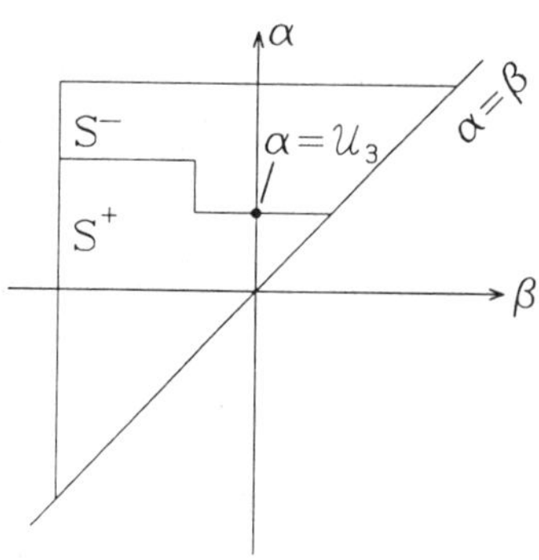

Fig. 1.10

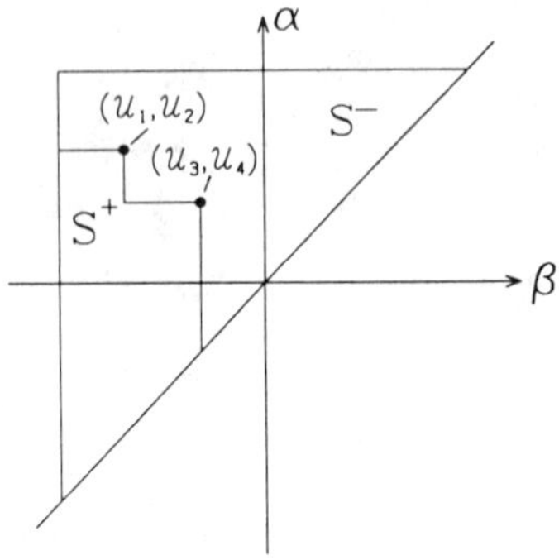

Fig. 1.11

in the "up" position, and $S^-(t)$ consisting of points (α, β) for which the corresponding $\hat{\gamma}$-operators are in the "down" position. The interface $L(t)$ between $S^+(t)$ and $S^-(t)$ is a staircase line whose vertices have α and β coordinates coinciding respectively with local maxima and minima of input at previous instants of time. The final link of $L(t)$ is attached to the line $\alpha = \beta$ and it moves when the input is changed. This link is a horizontal one and it moves up as the input is increased (see Fig. 1.12). The final link is a vertical one and it moves from right to left as the input is decreased (see Fig. 1.13).

According to the above conclusion, at any instant of time the integral in (1.1) can be subdivided into two integrals, over $S^+(t)$ and $S^-(t)$, respectively:

$$f(t) = \hat{\Gamma}u(t) = \iint\limits_{S^+(t)} \mu(\alpha, \beta)\hat{\gamma}_{\alpha\beta}u(t)\,d\alpha\,d\beta + \iint\limits_{S^-(t)} \mu(\alpha, \beta)\hat{\gamma}_{\alpha\beta}u(t)\,d\alpha\,d\beta. \quad (1.5)$$

Since

$$\hat{\gamma}_{\alpha\beta}u(t) = +1, \qquad \text{if } (\alpha, \beta) \in S^+(t) \quad (1.6)$$

and

$$\hat{\gamma}_{\alpha\beta}u(t) = -1, \qquad \text{if } (\alpha, \beta) \in S^-(t), \quad (1.7)$$

from (1.5) we find

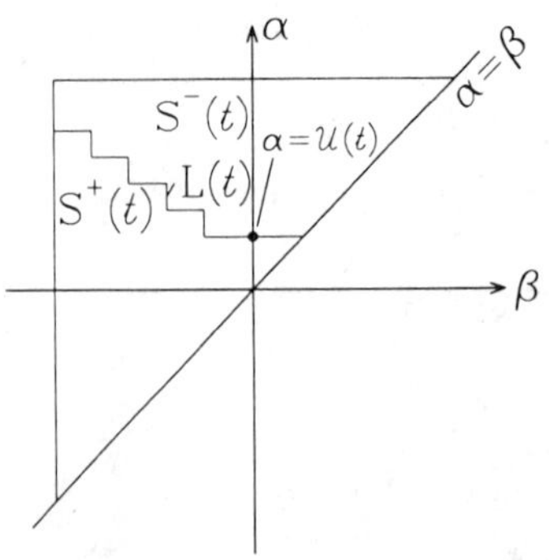

Fig. 1.12

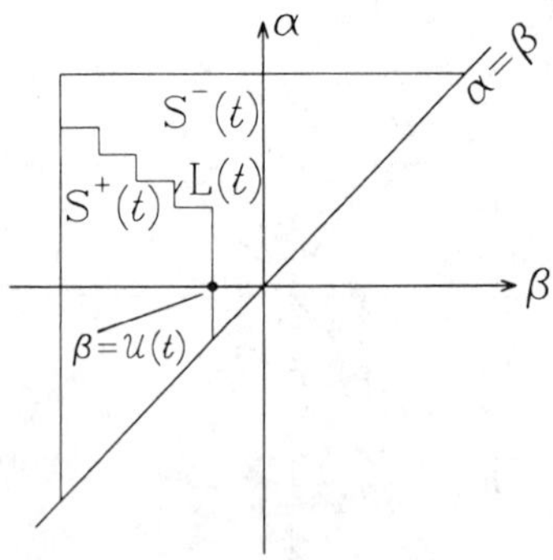

Fig. 1.13

$$f(t) = \iint\limits_{S^+(t)} \mu(\alpha, \beta) d\alpha d\beta - \iint\limits_{S^-(t)} \mu(\alpha, \beta) d\alpha d\beta. \tag{1.8}$$

From the above expression, it follows that an instantaneous value of output depends on a particular subdivision of the limiting triangle T into positive and negative sets $S^+(t)$ and $S^-(t)$. This subdivision is determined by a particular shape of the interface $L(t)$. This shape, in turn, depends on the past extremum values of input because these extremum values are the coordinates of the vertices of $L(t)$. Consequently, the past extremum values of input shape the staircase interface $L(t)$, and in this way they leave their mark upon the future.

To make the above point perfectly clear, consider two inputs $u_1(t)$ and $u_2(t)$ with two different past histories for $t < t'$. This means that they had different local extrema for $t < t'$. It is next assumed that these inputs coincide for $t \geq t'$. Then according to (1.8), the outputs $f_1(t)$ and $f_2(t)$ corresponding to the above inputs are given by the formulas

$$f_1(t) = \iint\limits_{S_1^+(t)} \mu(\alpha, \beta) d\alpha d\beta - \iint\limits_{S_1^-(t)} \mu(\alpha, \beta) d\alpha d\beta, \tag{1.9}$$

$$f_2(t) = \iint\limits_{S_2^+(t)} \mu(\alpha, \beta) d\alpha d\beta - \iint\limits_{S_2^-(t)} \mu(\alpha, \beta) d\alpha d\beta, \tag{1.10}$$

where $S_1^+(t)$ and $S_1^-(t)$, $S_2^+(t)$ and $S_2^-(t)$ are positive and negative sets of two subdivisions of T associated with $u_1(t)$ and $u_2(t)$, respectively.

The above two subdivisions are different because they correspond to two different input histories. Thus, from (1.9) and (1.10) we conclude

$$f_1(t) \neq f_2(t) \qquad \text{for } t > t'. \tag{1.11}$$

It is clear that inequality (1.11) holds even if the outputs $f_1(t')$ and $f_2(t')$ are somehow the same at time t'. This means that the Preisach model (1.1) describes, in general, hysteresis nonlinearities with nonlocal memories.

The above discussion reveals the mechanism of memory formation in the Preisach model. The memory is formed as a result of two different rules for the modification of the interface $L(t)$. Indeed, for a monotonically increasing input we have a horizontal final link of $L(t)$ moving upwards, while for a monotonically decreasing input we have a vertical final link of $L(t)$ moving from right to left. These two different rules result in the formation of the staircase interface $L(t)$, whose vertices have coordinates equal to past input extrema.

It is apparent from the previous analysis that the Preisach model can be defined in purely geometric terms, without any reference to the analytical definition (1.1). Indeed, the formula (1.8) along with the above two rules for the modificaton of $L(t)$ can be interpreted as an independent definition of the Preisach model. This definition is fully equivalent to the previous one. However, the geometric definition may be convenient for further generalization of the Preisach model. For instance, new and more general rules for the subdivision of T into positive and negative sets $S^+(t)$ and $S^-(t)$ may be introduced. In these rules, the links of $L(t)$ may not necessarily be the segments of straight lines parallel to coordinate axes. Furthermore, different functions $\mu^+(\alpha, \beta)$ and $\mu^-(\alpha, \beta)$ may be defined on the positive and negative sets, respectively. All these modifications may result in some meaningful generalizations of the Preisach model. However, the above possibilities have not been examined enough in the existing literature.

Having described the geometric interpretation of the Preisach model, we are now well equipped for the discussion of the main properties of this model. We begin with the simplest property which expresses the fact that the output value f^+ in the state of positive saturation is equal to the minus output value $-f^-$ in the state of negative saturation. In the state of positive saturation, the input $u(t)$ is more than α_0 and all $\hat{\gamma}$-operators are in the "up" position. Hence, according to (1.8), we find

$$f^+ = \iint_T \mu(\alpha, \beta)d\alpha d\beta. \tag{1.12}$$

Similarly, in the state of negative saturation the input $u(t)$ is less than β_0 and all $\hat{\gamma}$-operators are in the "down" position. As a result, we obtain

$$f^- = -\iint_T \mu(\alpha, \beta)d\alpha d\beta. \tag{1.13}$$

From (1.12) and (1.13), we have

$$f^+ = -f^-. \tag{1.14}$$

It is important to keep in mind that the saturation value f^+ and f^- remain constant for any value of input $u(t)$ above α_0 and below β_0, respectively. In other words, after ascending and descending branches merge together, they become flat. Partly for this reason, it is often said that the Preisach model does not describe reversible components of hysteresis nonlinearities. These components are regarded to be responsible for finite slopes of ascending and descending branches after they merge together. The inability of the Preisach model to describe the reversible components of hysteresis nonlinearities has long been viewed as a deficiency of the model. It will be shown later in the book that this deficiency along with some others can be removed by the appropriate generalization of the Preisach model.

We next proceed to the more interesting property which further elucidates the mechanism of memory formation in the Preisach model. It turns out that this model does not accumulate all past extremum values of input. Some of them can be wiped out by subsequent input variations. To make this property clear, consider a particular past history which is characterized by a finite decreasing sequence $\{u_1, u_3, u_5, u_7\}$ of local input maxima and an increasing sequence $\{u_2, u_4, u_6, u_8\}$ of local input minima. A typical $\alpha - \beta$ diagram for this kind of history is shown in Fig. 1.14. Now, we assume that the input $u(t)$

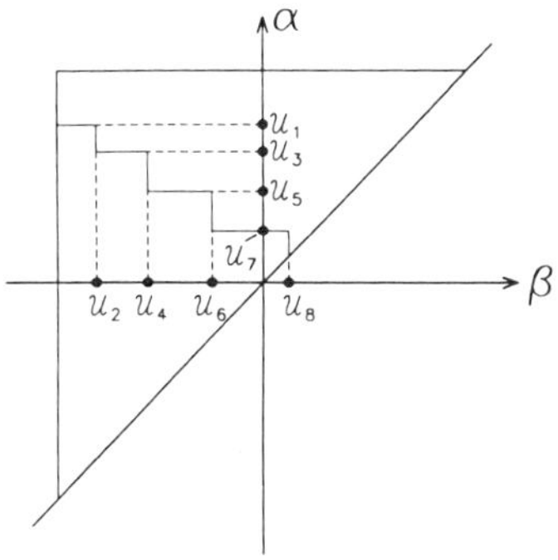

Fig. 1.14

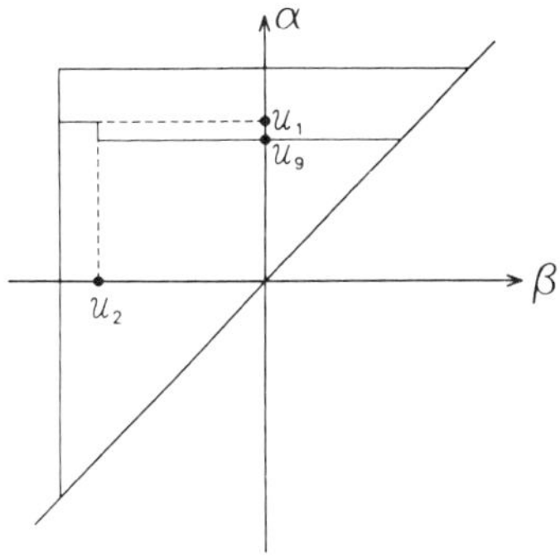

Fig. 1.15

is monotonically increased until it reaches some maximum value u_9 which is above u_3. This monotonic increase of inputs $u(t)$ results in the formation of a horizontal final link of $L(t)$ which moves upwards until the maximum value u_9 is reached. This results in a modified $\alpha - \beta$ diagram shown in Fig. 1.15. It is evident that all vertices whose α-coordinates were below u_9 have been wiped out. It is also clear that the wiping out of vertices is equivalent to the erasing the history associated with these vertices. Namely, the past input maxima and minima which were respectively equal to α- and β-coordinates of the erased vertices have been wiped out. We have illustrated how the wiping out of vertices occurs for monotonically increasing inputs. However, it is obvious that the wiping out of vertices occurs in a similar manner for montonically decreasing inputs as well. Thus, we can formulate the following property of the Preisach model.

Wiping-Out Property

Each local input maximum wipes out the vertices of $L(t)$ whose α-coordinates are below this maximum, and each local minimum wipes out the vertices whose β-coordinates are above this minimum.

The wiping-out property is asserted above in purely geometric terms. This makes this property quite transparent. However, the same property can also be described in analytical terms. The analytical formulation complements the geometric one because it is direclty phrased in terms of time input variations.

Consider a particular input variation shown in Fig. 1.16 for the time interval $t_0 \leq t \leq t'$. We assume that at the initial instant of time t_0 the input value $u(t_0)$ was below β_0. This means that the initial state is the state of negative saturation. Consequently, the whole history has been written by the input variation after time t_0. We would like to specify explicitly which local input extremum values will be stored by the Preisach model at time t'. Consider the global maximum of the input at the time interval $[t_0, t']$. We will use the notations M_1 for this maximum and t_1^+ for the instant of time the maximum was reached:

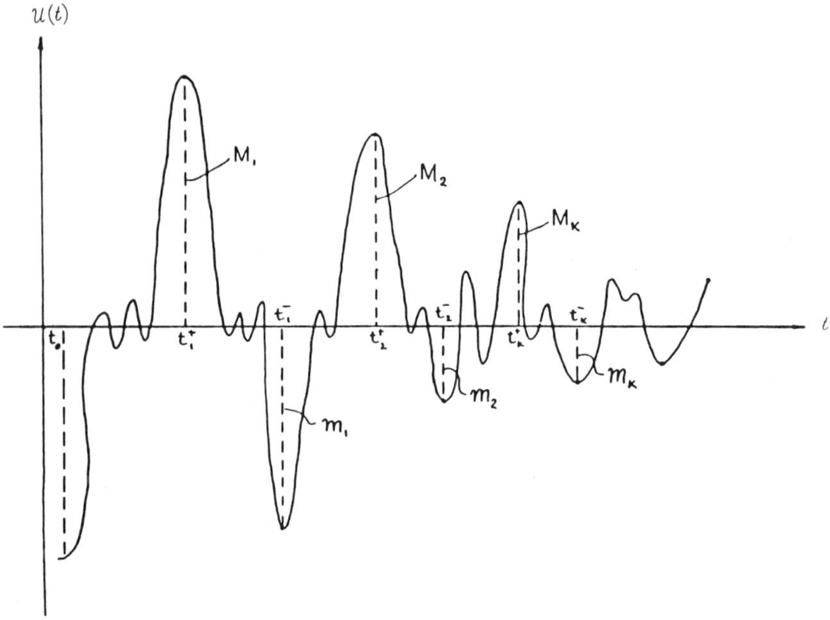

Fig. 1.16

$$M_1 = \max_{[t_0, t']} u(t), \qquad u(t_1^+) = M_1. \tag{1.15}$$

It is clear that all previous input extrema were wiped out by this maximum. Now, consider the global minimum of the input at the interval $[t_1^+, t']$. We will use the notation m_1 for this minimum and t_1^- for the time it was reached:

$$m_1 = \min_{[t_1^+, t']} u(t), \qquad u(t_1^-) = m_1. \tag{1.16}$$

It is apparent that all intermediate input extrema which occurred between t_1^+ and t_1^- were erased by the minimum m_1. Next, consider the global maximum of the input at the interval $[t_1^-, t']$. The notations M_2 and t_2^+ are appropriate for this maximum and the time it occurred:

$$M_2 = \max_{[t_1^-, t']} \omega(t), \qquad u(t_2^+) = M_2. \tag{1.17}$$

It is obvious that this maximum wiped out all intermediate input extrema which occurred between t_1^- and t_2^+. As before, consider the global minimum of input at the time interval $[t_2^+, t']$ and the notations m_2 and t_2^- will be used for this minimum and the time it was achieved:

$$m_2 = \min_{[t_2^+, t']} u(t), \qquad u(t_2^-) = m_2. \tag{1.18}$$

It is clear that this minimum wiped out all intermediate input extrema.

Continuing the above line of reasoning, we can inductively introduce the global maximum M_k and global minimum m_k:

$$M_k = \max_{[t_{k-1}^-, t']} u(t), \qquad u(t_k^+) = M_k \tag{1.19}$$

and

$$m_k = \min_{[t_k^+, t']} u(t), \qquad u(t_k^-) = m_k. \tag{1.20}$$

Only these input extrema were accumulated by the Preisach model, while all intermediate input extrema were erased. It is natural to say that M_k and m_k $(k = 1, 2, \ldots)$ form an alternating series of dominant global maxima and minima.

It is evident from the above analysis that α- and β-coordinates of vertices of the interface $L(t')$ are equal to M_k and m_k, respectively. It is also clear that the alternating series of dominant extrema is modified as time goes by. This means that new dominant extrema can be introduced by the time varying input, while the previous ones can be erased. In other words, M_k and m_k are functions of t' as it is clearly suggested by their definitions (1.19) and (1.20).

Now, the wiping-out property can be stated in the following form.

Wiping-Out Property

Only the alternating series of dominant input extrema are stored by the Preisach model. All other input extrema are wiped out.

It is worth noting that the wiping-out property is somewhat natural and consistent with experimental facts. Indeed, experiments in the area of magnetics show the existence of major hysteresis loops whose shapes do not depend on how these loops are approached. In other words, the major hysteresis loops are well defined. It means that any past history is wiped out by input oscillations of sufficiently large magnitude. This is in complete agreement with the wiping-out property.

Consider another characteristic property of the Preisach model which is valid for periodic input variations. Let $u_1(t)$ and $u_2(t)$ be two inputs which may have different past histories (different alternating series of dominant extrema). However, starting from some instant of time t_0, these inputs vary back and forth between the same two consecutive extremum values, u_+ and u_-. It can be shown that these periodic input variations result in minor hysteresis loops. Let Fig. 1.17 and 1.18 represent $\alpha - \beta$ diagrams for the inputs $u_1(t)$ and $u_2(t)$, respectively. As the inputs vary back and forth between u_+ and u_-, the final links of staircase interfaces $L_1(t)$ and $L_2(t)$ move within the identical triangles T_1 and T_2. This results in periodic shape variations for $L_1(t)$ and $L_2(t)$ which in turn produce periodic variations of the outputs $f_1(t)$ and $f_2(t)$. This means that some minor hysteresis loops are traced in the $f - u$ diagram for both inputs (see Fig. 1.19). The positions of these two loops with respect to the f-axis are different. This is because the above two inputs have different past histories which lead to different shapes for staircase interfaces $L_1(t)$ and $L_2(t)$. As a result, the values of outputs for the same values of inputs are different. This is easily seen from the formula (1.8). However, it can be proven that the above two hysteresis loops are congruent. It means that the

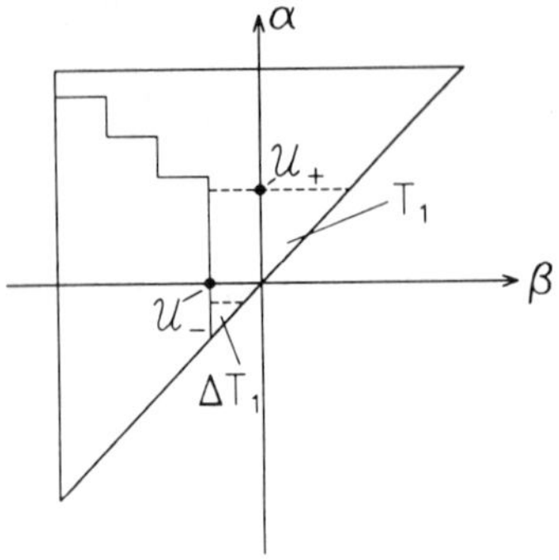

Fig. 1.17

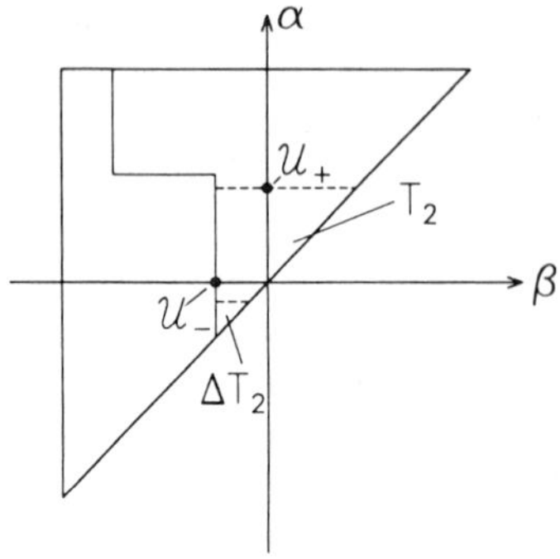

Fig. 1.18

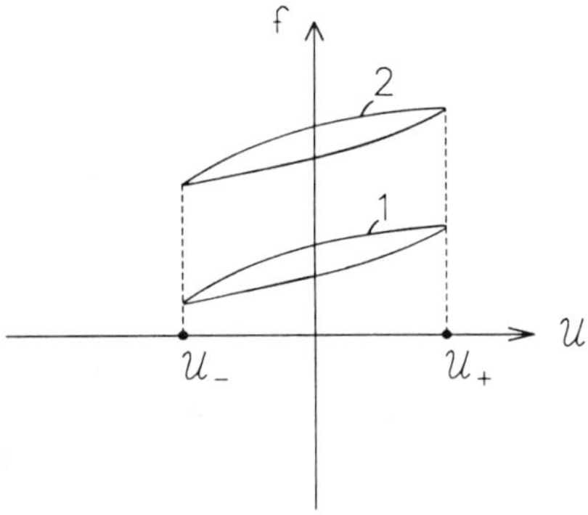

Fig. 1.19

coincidence of these loops can be achieved by the appropriate translation of these loops along the f-axis.

The proof of the congruency of the above loops is equivalent to showing that any equal increments of inputs $u_1(t)$ and $u_2(t)$ result in equal increments of outputs $f_1(t)$ and $f_2(t)$. To this end, let us assume that both inputs after achieving the same minimum value u_- are increased by the same amount: $\Delta u_1 = \Delta u_2 = \Delta u$. As a result of these increases, the identical triangles ΔT_1 and ΔT_2 are added to the positive sets $S_1^+(t)$ and $S_2^+(t)$ and subtracted from the negative sets $S_1^-(t)$ and $S_2^-(t)$ (see Fig. 1.17 and 1.18). Now, using the formula (1.8), we find that the corresponding output increments are given by

$$\Delta f_1 = 2 \iint_{\Delta T_1} \mu(\alpha, \beta) d\alpha d\beta, \tag{1.21}$$

$$\Delta f_2 = 2 \iint_{\Delta T_2} \mu(\alpha, \beta) d\alpha d\beta. \tag{1.22}$$

Since $\Delta T_1 = \Delta T_2$, we conclude

$$\Delta f_1 = \Delta f_2. \tag{1.23}$$

The equality (1.23) has been proven for the case when inputs $u_1(t)$ and $u_2(t)$ are monotonicaly increased by the same amount after achieving the same minimum value u_-. Thus, this equality means the congruency for the ascending branches of the above minor loops. By literally repeating the previous reasoning, we can prove that the same equality (1.23) holds when the inputs $u_1(t)$ and $u_2(t)$ are monotonically decreased by the same amount Δu after achieving the maximum value u_+. This means that the descending branches of the above minor loops are congruent as well. Thus, we have established the following property of the Preisach model.

Congruency Property

All minor hysteresis loops corresponding to back-and-forth variations of inputs between the same two consecutive extremum values are congruent.

Possible Applications of the Preisach Model as a neural network

We conclude this paragraph with a remark on possible applications of the Preisach model beyond the area of hysteresis modeling. It is clear from the previous discussion that the Preisach model has the ability to detect and store the alternating series of dominant extrema of input. In other words, the Preisach model is endowed with memory. For this reason, the Preisach model might have appeal as the mathematical model of memory with some interesting properties, and its device realization (see Fig. 1.2) might be utilized as an unusual storage device. We will discuss below only a few peculiarities of this memory.

First, the mechanism of memory formation in the Preisach model is quite simple and results from the parallel connection of qualitatively similar elements (two-position cells) $\hat{\gamma}_{\alpha\beta}$. Since this model employs very simple elements and has little structure, its memory formation can be interpreted as a "spontaneously emerged collective effect." (This kind of terminology is in vogue nowadays in the area of neural networks; see, for instance, [36]).

Second, storage of information is not localized. Indeed, the model (1.1) stores the informaton (extremum values of input) not in particular separate cells (as in the case of computer storage devices), but some ensembles of the cells $\hat{\gamma}_{\alpha\beta}$ participate in storage of each bit of information. As a result, if some of the cells are destroyed, the stored information might still be preserved. This opens the way for cell replenishment without affecting the currently stored information.

Third, some subsequent events may erase the information previously stored in the memory (wiping-out property). This erasing occurs if only new events make stronger impact (characterized by larger input extrema) than previous ones.

Fourth, as it will be shown in Section 6, the Preisach model exhibits "fading" memory. This means that the stored information is slowly and gradually erased in the presence of small random perturbations.

The above properties are somewhat similar to those being observed (or suspected) for memories in biological systems. However, it will be imprudent to speculate how far this similarity goes. Nevertheless, it may be expected that the mathematical tool (1.1) might find some interesting applications beyond the conventional area of hysteresis modeling.

3 Experimental Determination of $\mu(\alpha, \beta)$. Representation Theorem

To determine $\mu(\alpha, \beta)$, the set of first-order transition (reversal) curves is needed. These curves can be experimentally found as follows. First, the input $u(t)$ should be decreased to the value which is less than β_0. It brings a hysteresis transducer to the state of negative saturation. Next, the input is monotonically increased until it reaches some value α'. The corresponding $\alpha - \beta$ diagram is shown in Fig. 1.20. As the input is increased, an ascending branch of a major loop is followed (see Fig. 1.21). This branch will be also called the limiting ascending branch because usually there is no branch below it. The notation $f_{\alpha'}$ will be used for the output value on this branch which corresponds to the input value $u = \alpha'$. The first-order transition (reversal) curves are attached to the limiting ascending branch. Each of these curves is formed as the above monotonic increase of the input is followed by a subsequent monotonic decrease. The term "first-order" is used to emphasize the fact that each of these curves is formed after the first reversal of input. The notation $f_{\alpha'\beta'}$ will be used for the output value on the transition curve attached to the limiting ascending branch at the point $f_{\alpha'}$. This output value corresponds to the input value $u = \beta'$ (see Fig. 1.21). The above monotonic decrease of input modifies the previous $\alpha - \beta$ diagram shown in Fig. 1.20. A new $\alpha - \beta$ diagram for the instant of time when the input reaches the value β' is illustrated by Fig. 1.22.

Now, we define the function

$$F(\alpha', \beta') = \tfrac{1}{2}(f_{\alpha'} - f_{\alpha'\beta'}). \qquad (1.24)$$

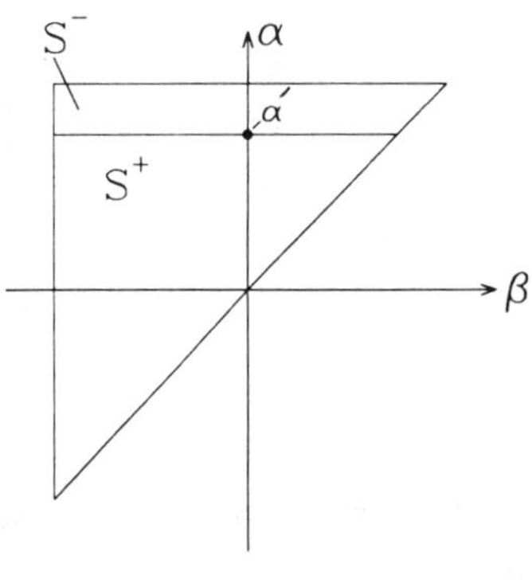

Fig. 1.20

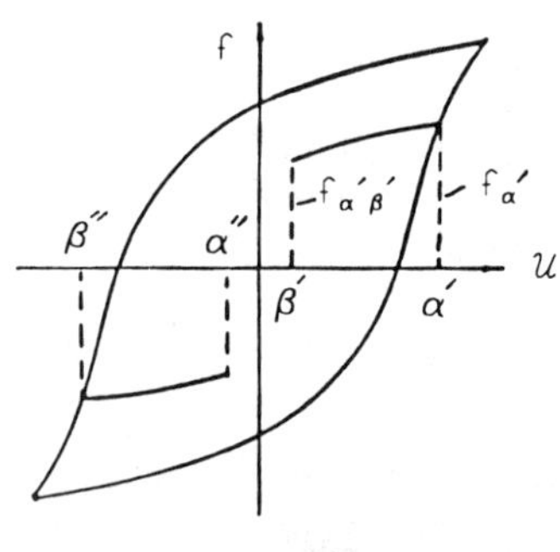

Fig. 1.21

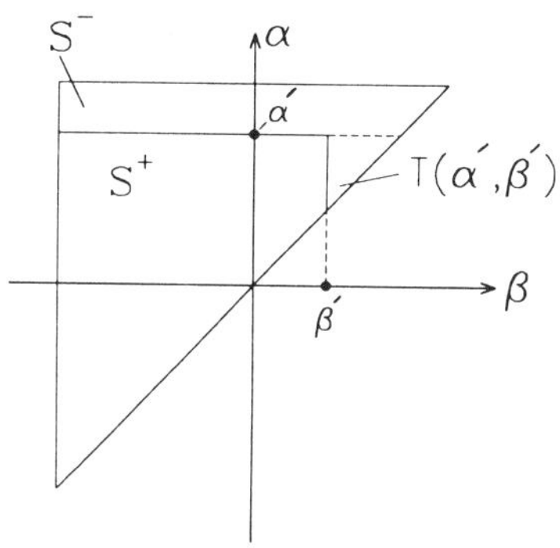

Fig. 1.22

This function is equal to one-half of the output increments along the first-order transition curves. The next step is to express this function in terms of the Preisach function $\mu(\alpha, \beta)$. To this end, we compare the $\alpha - \beta$ diagrams shown in Figs. 1.20 and 1.22. It is clear from these diagrams that the triangle $T(\alpha', \beta')$ is added to the negative set S^- and subtracted from the positive set S^+ as a result of the monotonic input decrease from the value $u = \alpha'$ to the value $u = \beta'$. Using the above fact and the formula (1.8), we find that the Preisach model will match the output increments along the first-order transition curves if the function $\mu(\alpha, \beta)$ satisfies the equation

$$f_{\alpha'\beta'} - f_{\alpha'} = -2 \iint\limits_{T(\alpha', \beta')} \mu(\alpha, \beta)\, d\alpha\, d\beta. \tag{1.25}$$

By comparing the formulas (1.24) and (1.25), we obtain

$$F(\alpha', \beta') = \iint\limits_{T(\alpha', \beta')} \mu(\alpha, \beta)\, d\alpha\, d\beta. \tag{1.26}$$

The integral over the triangle $T(\alpha', \beta')$ can be written as the following double integral:

$$F(\alpha', \beta') = \int_{\beta'}^{\alpha'} \left(\int_{\beta}^{\alpha'} \mu(\alpha, \beta)\, d\alpha \right) d\beta. \tag{1.27}$$

By differentiating the last expression twice (first with respect to β' and then with respect to α'), we find

$$\mu(\alpha', \beta') = -\frac{\partial^2 F(\alpha', \beta')}{\partial \alpha' \partial \beta'}. \tag{1.28}$$

Invoking (1.24), the expression (1.28) can be written in another equivalent form:

$$\mu(\alpha', \beta') = \frac{1}{2} \frac{\partial^2 f_{\alpha'\beta'}}{\partial \alpha' \partial \beta'}. \tag{1.29}$$

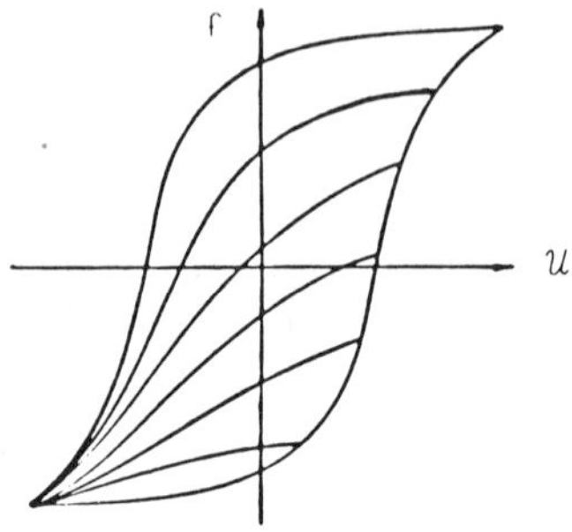

Fig. 1.23

The formula (1.29) allows for a simple geometric interpretation of the function $\mu(\alpha', \beta')$. Indeed, for the first derivative of $f_{\alpha'\beta'}$ with respect to β' we have

$$\frac{\partial f_{\alpha'\beta'}}{\partial \beta'} = \tan \theta(\alpha', \beta'), \tag{1.30}$$

where $\theta(\alpha', \beta')$ is the angle between the axis u and the tangent to the first-order transition curve $f_{\alpha'\beta'}$ at the point $u = \beta'$.

From (1.29) and (1.30), we find

$$\mu(\alpha', \beta') = \frac{1}{2} \frac{\partial \tan \theta(\alpha', \beta')}{\partial \alpha'}. \tag{1.31}$$

From (1.31), we conclude that the Preisach function $\mu(\alpha', \beta')$ is positive if $\tan \theta(\alpha', \beta')$ is a monotonicaly increasing function of α' for any fixed β'. The last condition is satisfied if all first-order transition curves are monotonically increasing functions of β', and they do not intersect inside the major loop but merge together at the point where the descending and ascending branches of the major loop meet one another (see Fig. 1.23).* To secure the above merge of first-order transition curves, $\tan \theta(\alpha', \beta')$ should increase as a function of α' for any fixed β' to compensate for larger values of $f_{\alpha'\beta'}$. It is worthwhile noting that in the "magnetic" definition of the Preisach model (see Section 1) the function $\mu(\alpha, \beta)$ is treated as a distribution function with some statistical connotation. For this reason, this function is tacitly assumed to be positive. In our treatment, the last property is directly related to experimental facts.

For positive $\mu(\alpha, \beta)$, $F(\alpha', \beta')$ is a monotonically increasing function of α' for any fixed β' and a monotonically decreasing function of β' for any fixed α'. This follows directly from (1.26).

If the Preisach function is determined from (1.29) (or (1.28)), then the expression (1.25) is satisfied. This means that the Preisach model matches the output increments along the first-order transition curves. We next show that

* It is useful to keep in mind that the descending branch of the major loop can be interpreted as a particular case of the first-order transition curve for which $\alpha' = \alpha_0$.

this also implies that the first-order transition curves themselves are matched. To this end, consider the particular case when $\beta' = \beta_0$. For this case, we have

$$f_{\alpha'\beta_0} = f^-. \tag{1.32}$$

From (1.25) and (1.32), we obtain

$$f^- - f_{\alpha'} = -2 \iint_{T(\alpha', \beta_0)} \mu(\alpha, \beta) d\alpha d\beta. \tag{1.33}$$

Thus, the Preisach model matches the output increment, $f^- - f_{\alpha'}$, along the limiting ascending branch.

For $\alpha' = \alpha_0$, we have

$$f_{\alpha_0} = f^+. \tag{1.34}$$

From (1.34) and (1.33), we find

$$f^- - f^+ = -2 \iint_T \mu(\alpha, \beta) d\alpha d\beta, \tag{1.35}$$

where $T = T(\alpha_0, \beta_0)$ is the limiting triangle.

Since major hysteresis loops are usually symmetric, the following equality holds:

$$f^+ = -f^-. \tag{1.36}$$

From (1.35) and (1.36), we derive

$$f^- = -\iint_T \mu(\alpha, \beta) d\alpha d\beta. \tag{1.37}$$

This means that the Preisach model matches the output value in the state of negative saturation. From the last fact and (1.33), we conclude that the model matches the limiting ascending branch. Since the limiting ascending branch and the output increments along first-order transition curves are matched, we deduce that the first-order transition curves themselves are matched.

The Preisach function, $\mu(\alpha, \beta)$, has been found by using the first-order transition curves. These curves are attached to the limiting ascending branch, and each of them is formed when a monotonic increase along this branch is followed by a subsequent input decrease. For this reason, these curves can be named first-order *decreasing* transition curves. However, by almost literally repeating the previous reasoning, a similar expression for $\mu(\alpha, \beta)$ can be found by using the first-order *increasing* transition curves. These curves are attached to the descending branch of the major loop. Each of these first-order increasing transition curves is formed as a monotonic decrease along the limiting descending branch is followed by a subsequent input increase. The notation $f_{\beta''}$ will be used for the output value on the limiting descending branch. This value is achieved when the input is monotonically decreased from some value above

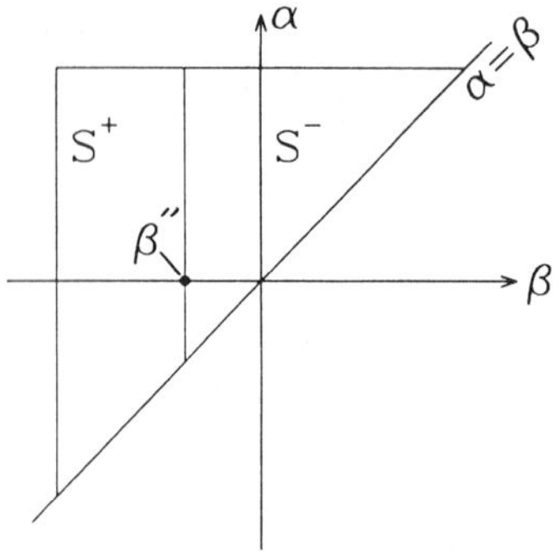

Fig. 1.24

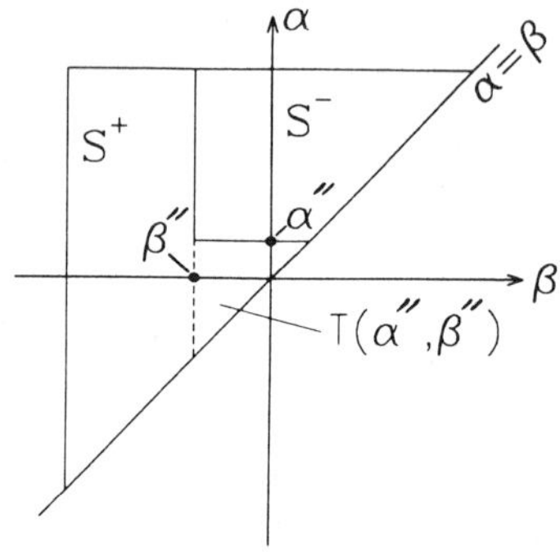

Fig. 1.25

α_0 to the value $u = \beta''$. The corresponding $\alpha - \beta$ diagram is shown in Fig. 1.24. The notation $f_{\beta''\alpha''}$ will be used for the output value on the first-order increasing transition curve which is attached to the limiting descending branch at the point $f_{\beta''}$. This output value corresponds to $u = \alpha''$ (see Fig. 1.21). The corresponding $\alpha - \beta$ diagram is shown in Fig. 1.25.

By using the function

$$F(\alpha'', \beta'') = \iint_{T(\alpha'', \beta'')} \mu(\alpha, \beta) d\alpha d\beta, \tag{1.38}$$

from Fig. 1.24 and 1.25 and the formula (1.8), we obtain

$$F(\alpha'', \beta'') = \tfrac{1}{2}(f_{\beta''\alpha''} - f_{\beta''}). \tag{1.39}$$

From (1.38), as before, we find

$$\mu(\alpha'', \beta'') = -\frac{\partial^2 F(\alpha'', \beta'')}{\partial \alpha'' \partial \beta''}. \tag{1.40}$$

It is clear on the physical grounds (symmetry considerations) that first-order decreasing and increasing transition curves are congruent. In mathematical terms, this means that if

$$\beta'' = -\alpha', \quad \text{and} \quad \alpha'' = -\beta'', \tag{1.41}$$

then

$$f_{\beta''} = -f_{\alpha'} \quad \text{and} \quad f_{\beta''\alpha''} = -f_{\alpha'\beta'}. \tag{1.42}$$

From (1.24), (1.39), and (1.42), we find that if (1.41) holds, then

$$F(\alpha'', \beta'') = F(\alpha', \beta'). \tag{1.43}$$

By substituting (1.41) into (1.43), we obtain

$$F(-\beta', -\alpha') = F(\alpha', \beta'). \tag{1.44}$$

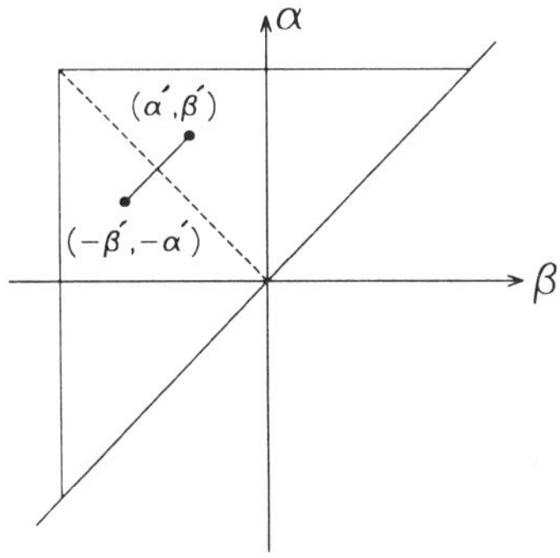

Fig. 1.26

Next, substituting (1.43) and (1.41) into (1.40), we derive

$$\mu(-\beta', -\alpha') = -\frac{\partial^2 F(\alpha', \beta')}{\partial \beta' \partial \alpha'}. \tag{1.45}$$

By comparing (1.45) and (1.28), we find

$$\mu(-\beta', -\alpha') = \mu(\alpha', \beta'). \tag{1.46}$$

The formulas (1.44) and (1.46) express the mirror symmetry of functions $F(\alpha, \beta)$ and $\mu(\alpha, \beta)$ with respect to the line $\alpha = -\beta$ (see Fig. 1.26). This symmetry is a consequence of the congruency of the first-order decreasing and increasing transition curves.

If the line $\alpha = -\beta$ is the interface between the positive and negative sets $S^+(t)$ and $S^-(t)$, then, according to the above symmetry and (1.8), we find that the output is equal to zero:

$$f(t) = 0. \tag{1.47}$$

For this reason, the state corresponding to the above interface is called in magnetics "the demagnetized state." However, this state can not be exactly

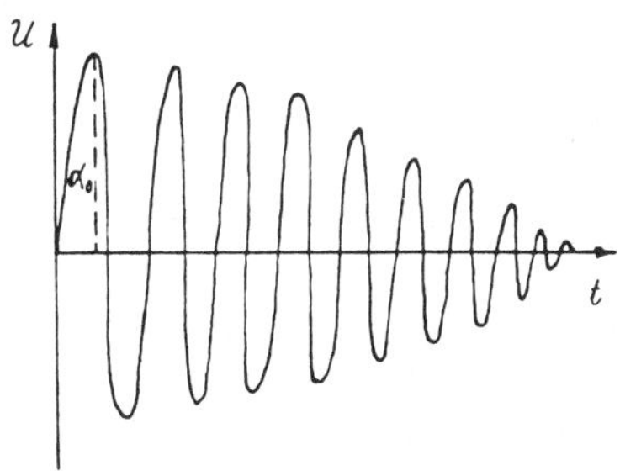

Fig. 1.27

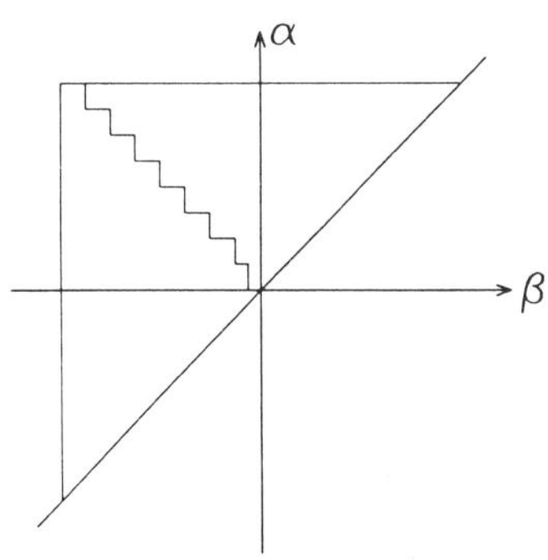

Fig. 1.28

achieved. This is because an interface $L(t)$ is always a staircase line whose links are parallel to α and β axes. As a result, an actual staircase interface can only approximate the line $\alpha = -\beta$. Such an approximation can be achieved, for instance, if the input $u(t)$ is an oscillating function whose amplitude is slowly decreased to zero starting from some value above α_0 (see Fig. 1.27). The corresponding staircase interface is shown in Fig. 1.28.

We next proceed to the formulation and the proof of the fundamental theorem which gives the necessary and sufficient conditions for the representation of actual hysteresis nonlinearities by the Preisach model.

Representation Theorem

The wiping-out property and the congruency property constitute the necessary and sufficient conditions for a hysteresis nonlinearity to be represented by the Preisach model on the set of piecewise monotonic inputs.

Proof

Necessity

Let a hysteresis transducer be representable by the Preisach model. Then, this transducer should have the same properties as the model. In particular, it should have the wiping-out and congruency properties.

Sufficiency

Consider a hysteresis transducer which has both the wiping-out property and the congruency property. It can be recalled that the wiping-out property means that the future values of output do not depend on all past extremum values of input but only on those which form an alternating series of dominant input extrema. The congruency property, on the other hand, suggests that all minor hysteresis loops formed as a result of back-and-forth input variations between the same two consecutive input extrema are congruent. We intend to prove that the above properties imply that the hysteresis transducer can be represented by the Preisach model.

The proof of the last statement is constructive. First, it is assumed that the weight function $\mu(\alpha, \beta)$ is found for the given transducer by matching its first-order transition curves. This can be accomplished by using the formula (1.28). This formula is equivalent to (1.25) which, in turn, means that the integrals of $\mu(\alpha, \beta)$ over triangles $T(\alpha', \beta')$ are equal to one-half of output increments $\frac{1}{2}\Delta f = \frac{1}{2}(f_{\alpha'} - f_{\alpha'\beta'})$ along the first-order transition curves. Next, it will be proven that if the above weight function is substituted in (1.1), then the Preisach model and the given transducer will have the same input-output relationships provided that the wiping-out and congruency properties hold for the transducer. This statement is true for the first-order transition curves due to the very way by which the weight function $\mu(\alpha, \beta)$ is determined. The induction argument will be used to prove that the same statement holds for higher-order transition curves as well. Let us assume that the above statement

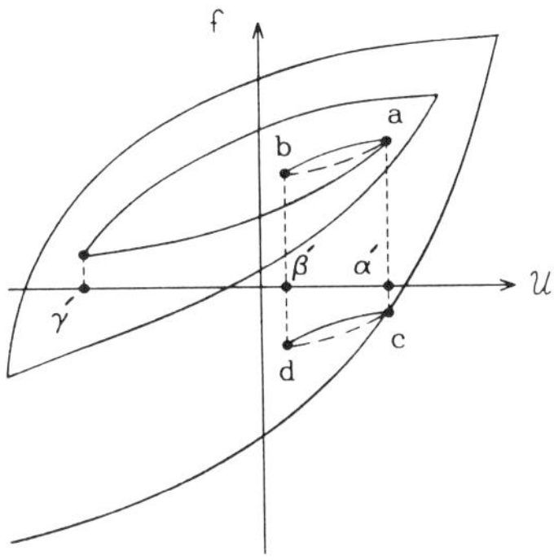

Fig. 1.29

is true for transition curves with number 1, 2, ... k. Then, for the induction inference, we need to prove that this statement holds for a transition curve number $k + 1$.

Let a be a point at which the transition curve number $k + 1$ starts. This is shown in Fig. 1.29, which represents the input-output relationship for the transducer. The point a corresponds to some input value $u = \alpha'$. According to the induction assumption, the output values of the transducer and the Preisach model coincide at this point. Thus, it remains to be proven that the output increments along the transition curve number $k + 1$ are the same for the actual transducer and for the Preisach model. Consider an arbitrary input value $u = \beta' < \alpha'$. The output increment for the transducer will be equal to the increment of f along some curve ab (see Fig. 1.29). Let Fig. 1.30 and 1.31 represent $\alpha - \beta$ diagrams for the Preisach model at the time instants when $u = \alpha'$ and $u = \beta'$, respectively. From these diagrams we find that the input decrease from α' to β' results in adding the triangle $T(\alpha', \beta')$ to the negative set S^- and subtracting the same triangle from the positive set S^+. Using the above fact and the formula (1.8), we find that for the Preisach model the output increment along the transition curve number $k + 1$ is given by

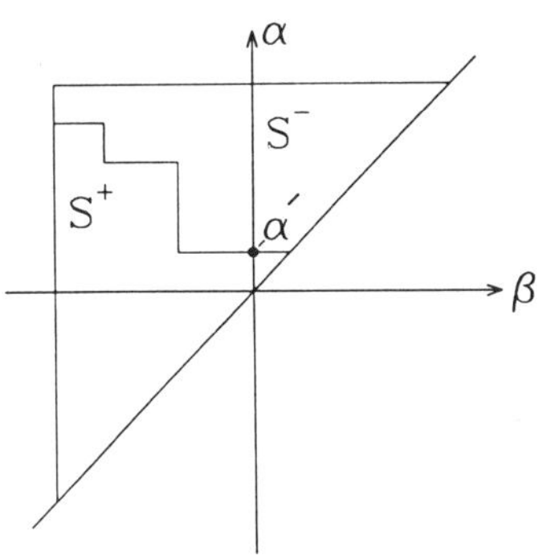

Fig. 1.30

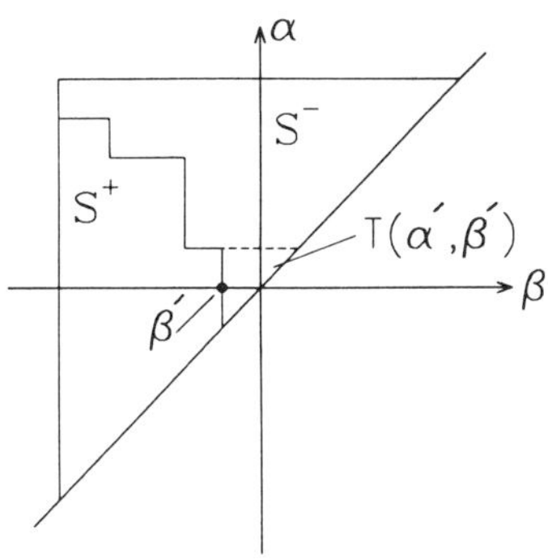

Fig. 1.31

$$\Delta f = 2 \iint_{T(\alpha', \beta')} \mu(\alpha, \beta) d\alpha d\beta. \tag{1.48}$$

However, according to the way the function $\mu(\alpha, \beta)$ is defined, the right-hand side of (1.48) is equal to the increment of the transducer output along the first-order transition curve *cd* (see Fig. 1.29). Thus, it remains to be shown that the output increments along the curves *ab* and *cd* are the same. It is here that the wiping-out and congruency properties will be used. The proof proceeds as follows. If starting from the point *b* we monotonically increased the input value from β' back to α', then, according to the wiping-out property, we would arrive at the same point *a* by moving along some curve *ba* (see Fig. 1.29). Indeed, the wiping-out property implies that as soon as the input exceeded the value α', the history associated with back-and-forth input variations between α' and β' should be erased and the subsequent output variation should follow the transition curve number *k*. But, this would be possible if only for $u = \alpha'$ we arrived back at the point *a*. Similarly, if starting from the point *d* we monotonically increased the input value from β' to α', then, according to the same wiping-out property, we would arrive at the point *c* moving along some curve *dc* (see Fig. 1.29). Now, invoking the congruency property, we conclude that the hysteresis loops *bab* and *dcd* are congruent. This is true because both loops result from back and forth input variations between the same two consecutive input extrema, α' and β'. From the congruency of the above loops, we find that the output increments along the curves *ab* and *cd* are the same. Consequently, the output values for the transducer and for the Preisach model coincide along the transition curve number $k + 1$. The last fact has been proven for any $\beta' \geq \gamma'$ (see Fig. 1.29). However, according to the wiping-out property, the transition curve number $k + 1$ should coincide with the transition curve number $k - 1$ for $\beta' < \gamma'$. Thus, the case $\beta' < \gamma'$ falls within the domain of the induction assumption.

In the above discussion, we have considered the transition curve number $k + 1$ corresponding to the monotonically decreasing input. However, the case when the transition curve number $k + 1$ is formed as a result of monotonically increasing input can be treated similarly. The only difference in the proof will be that the first-order *increasing* transition curves must be employed instead of the first-order *decreasing* transition curves used in the above reasoning. The details of the proof for the latter case are left to the reader as a useful exercise.

This completes the proof of the theorem.

Discussion

It is easy to see that the essence of the given proof is in the reduction of higher-order transition curves to the first-order transition curves. This reduction rests on both the wiping-out property and the congruency property. It also shows that any higher-order transition curve is congruent with some particular first-order transition curve which starts at the same reversal value of input α' as the above higher-order transition curve.

The proven theorem is important because it clearly establishes the limits of applicability of the Preisach model. These limits are formulated in purely phenomenological terms, without any reference to the actual physical nature of hysteresis. The theorem also explicitly indicates the factors which affect the accuracy of the Preisach model. Indeed, deviations from the conditions of the representation theorem may serve as the measure of model accuracy.

The above theorem allows for a simple explanation of the so-called statistical instability of the Preisach model. This instability is a very popular topic in the "magnetic" literature. It usually means that weight functions $\mu(\alpha, \beta)$ determined from different experimental data are not identical. Since these functions have been construed in magnetics as "particle distributions" (with some statistical connotation), the dependence of $\mu(\alpha, \beta)$ on a particular experimental way of its determination has been termed as "statistical instability." The origin of this "statistical instability" can be easily understood from the following discussion.

The formula (1.28) for $\mu(\alpha, \beta)$ has been derived from the expression (1.26). This means that the weight function $\mu(\alpha, \beta)$ can be always determined if the integrals of $\mu(\alpha, \beta)$ over the triangles $T(\alpha', \beta')$ are somehow experimentally found. It has already been proposed (see (1.25)) to find these integrals by matching the first-order transition curves. However, this is not the only way it can be done. Indeed, according to the formula (1.48), the same integrals can be found by matching the output increments along any particular transition curves. But, we will end up with the same values for these integrals if and only if higher-order transition curves and the first-order transition curves are congruent. According to the proof of the representation theorem, this congruency takes place if and only if the wiping-out and congruency properties are valid for hysteresis transducers. If these properties are not valid, then, by matching different transition curves, we will end up with different values for the integrals over the triangles $T(\alpha', \beta')$ and, consequently, with different values of $\mu(\alpha, \beta)$. This is exactly what happened when the "statistical instability" was discovered (see [18]). Thus, from the phenomenological point of view, the origin of "statistical instability" comes from the fact that the wiping-out and congruency properties may not be exactly satisfied for actual hysteresis nonlinearities. Under these circumstances, the Preisach model can not serve as an absolutely accurate representation for actual hysteresis nonlinearities, but it can still be used as an approximation. The quality of this approximation will depend on the extent to which the wiping-out and congruency properties are satisfied.

It has already been mentioned that the weight function $\mu(\alpha, \beta)$ can be found by matching output increments along any transition curves. Thus, the question arises: "Which transition curves should be used for the above purpose?" If the wiping-out and congruency properties are valid, then it really does not matter which transition curves are used for the determination of $\mu(\alpha, \beta)$. All of them will theoretically lead to the same result. However, from the practical point of view, the first-order transition curves have some clear advantages. First, it is easier to find these curves experimentally than higher-order transition curves. Second, measurements of these curves start from a well-defined state, namely, the state of negative (or positive) saturation. This is not the case for some experimental techniques described in the literature. For instance, it has been suggested in the literature to use a demagnetized state as a starting state in the experimental determination of $\mu(\alpha, \beta)$. But, as was pointed out before, the demagnetized state is not well defined for the Preisach model. This state depends on a particular way it has been prepared. As a result, some errors and discrepancies may be introduced if the demagnetized state is used as the initial state for the experimental determination of $\mu(\alpha, \beta)$.

In the case when the wiping-out and congruency properties are not valid, the values of $\mu(\alpha, \beta)$ and the accuracy of the Preisach model will depend on a particular choice of transition curves employed for the determination of $\mu(\alpha, \beta)$. A general recipe in this situation is to predict (or accuractely guess) possible regimes in which the hysteresis transducer will be used. Then, $\mu(\alpha, \beta)$ can be found from matching the transition curves which are expected to be followed in these regimes.

It is often asserted in the literature that the Preisach model describes reversal curves with zero initial slopes. This statement is based on the following reasoning. Let $f_{\alpha'\beta'}$ be a reversal curve which is traced for a monotonically decreasing input. This curve starts at the point $f_{\alpha'}$ which corresponds to the local input maximum $u = \alpha'$. The difference $f_{\alpha'\beta'} - f_{\alpha'}$ is then given by (1.25) and the current slope of this reversal curve at the point $\beta = \beta'$ can be found from (1.30). By using (1.25), (1.26), (1.27), and (1.30), it is easy to derive the following expression for the initial slope

$$\lim_{\beta' \to \alpha'} \tan \theta(\alpha', \beta') = 2 \lim_{\beta' \to \alpha'} \int_{\beta'}^{\alpha'} \mu(\alpha, \beta') d\alpha.$$

A similar expression is valid for an initial slope of a reversal curve $f_{\beta'\alpha'}$ which is traced for a monotonically increasing input. This expression is given by

$$\lim_{\alpha' \to \beta'} \tan \theta(\alpha', \beta') = 2 \lim_{\alpha' \to \beta'} \int_{\beta'}^{\alpha'} \mu(\alpha', \beta) d\beta.$$

In the literature, the function $\mu(\alpha, \beta)$ is usually construed as a distribution function for magnetic particles and, for this reason, it is tacitly assumed to be bounded. This assumption and the last two formulas result in zero initial slopes for the reversal curves $f_{\alpha'\beta'}$ and $f_{\beta'\alpha'}$. This property is often regarded as an intrinsic property of the Preisach model. However, this is not true. Indeed, if the phenomenological approach is adopted, then there is no need to interpret μ as a distribution function for magnetic particles, and the assumption that μ is bounded can be removed. The freedom of choice of function $\mu(\alpha, \beta)$ should be used in order to match as many experimental facts as possible. It is easy to show that experimentally observed nonzero initial slopes of reversal curves $f_{\alpha'\beta'}$ and $f_{\beta'\alpha'}$ can be matched by the Preisach model if we allow for delta-type singularities of the function μ. Indeed, let the function μ be represented as

$$\mu(\alpha, \beta) = \tilde{\mu}(\alpha, \beta) + k(\alpha, \beta)\delta(\alpha - \beta),$$

where $\tilde{\mu}$ is a bounded function, and $\delta(\alpha - \beta)$ is the Dirac delta function. Then, from the last three formulas we find

$$\lim_{\beta' \to \alpha'} \tan \theta(\alpha', \beta') = 2 \lim_{\beta' \to \alpha'} \lim_{\alpha \to \beta'} k(\alpha, \beta')$$

and

$$\lim_{\alpha' \to \beta'} \tan \theta(\alpha', \beta') = 2 \lim_{\alpha' \to \beta'} \lim_{\beta \to \alpha'} k(\alpha', \beta).$$

Thus, if the function $k(\alpha, \beta)$ has different directional limits at the points of the line $\alpha = \beta$, then nonzero initial slopes of the reversal curves can be matched. It can also be shown that the above delta-type singularities guarantee the proper solution of equation (1.26). Without specifying these singularities, expression (1.28) can be regarded as a solution of equation (1.26) with the left-hand side $f(\alpha', \beta') + \varphi(\alpha') + \varphi(\beta')$ where $\varphi(\alpha')$ and $\varphi(\beta')$ are arbitrary functions.

This section will be concluded by the discussion of a simple example which illustrates the application of the formula (1.29). Consider a hysteresis nonlinearity with a local memory shown in Fig. 1.32. This hysteresis nonlinearity has a set of inner curves which are parallel to the u-axis. The above hysteresis nonlinearity is completely characterized by the following parameters: m_s, α_1, α_2, w. Our goal is to find the Preisach representation (1.1) for this nonlinearity. To this end, we have to find the weight function $\mu(\alpha, \beta)$. This requires the specification of first-order transition curves $f_{\alpha\beta}$ and the subsequent application of the formula (1.29). It is clear from Fig. 1.32 that for any α such that $\alpha_1 < \alpha < \infty$ the corresponding first-order transition curve $f_{\alpha\beta}$ consists of two parts: (a) a particular inner curves and (b) a part of the limiting descending branch if $\alpha_1 - w \leq \beta \leq \alpha_2 - w$, $\beta + w \leq \alpha < \infty$. It is apparent that $f_{\alpha\beta}$ varies only along the second part.

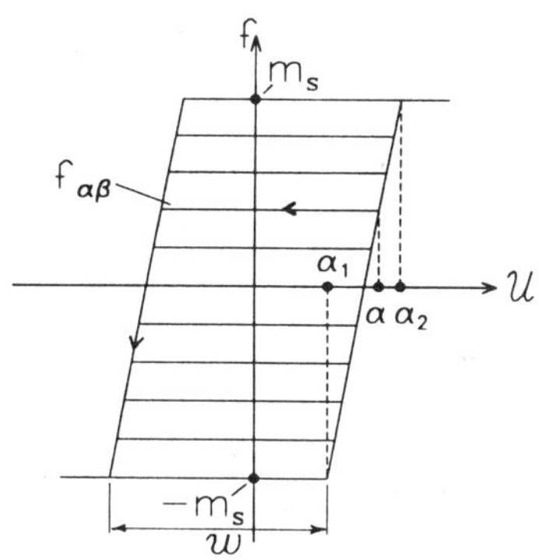

Fig. 1.32

From the above remarks, we find

$$f_{\alpha\beta} = -m_s + \frac{2m_s}{\alpha_2 - \alpha_1}(\beta - \alpha_1 + w), \tag{1.49}$$

if

$$\alpha_1 - w \leq \beta \leq \alpha_2 - w, \qquad \beta + w \leq \alpha \leq \infty, \tag{1.50}$$

and

$$f_{\alpha\beta} = \text{const} \tag{1.51}$$

for all other possible values of β.

It is worthwhile to keep in mind that, depending on a particular value of β, the constant in (1.51) may assume one of the following two values: $-m_s$, $-m_s + 2m_s(\alpha - \alpha_1)/(\alpha_2 - \alpha_1)$.

From (1.49), (1.50), and (1.51), we derive

$$\frac{\partial f_{\alpha\beta}}{\partial\beta} = \frac{2m_s}{\alpha_2 - \alpha_1}, \tag{1.52}$$

if

$$\alpha_1 - w \leq \beta \leq \alpha_2 - w, \qquad \beta + w \leq \alpha \leq \infty, \tag{1.53}$$

and

$$\frac{\partial f_{\alpha\beta}}{\partial\beta} = 0 \tag{1.54}$$

for all other values of α and β. The last three formulas can be written in the following equivalent form:

$$\frac{\partial f_{\alpha\beta}}{\partial\beta} = \frac{2m_s}{\alpha_2 - \alpha_1}, \qquad \text{if } (\alpha, \beta) \in \Omega, \tag{1.55}$$

and

$$\frac{\partial f_{\alpha\beta}}{\partial\beta} = 0, \qquad \text{if } (\alpha, \beta) \notin \Omega, \tag{1.56}$$

where Ω is the region defined by the inequalities (1.53) (see also Fig. 1.33).

From (1.55) and (1.56), we obtain

$$\frac{\partial^2 f_{\alpha\beta}}{\partial\alpha\partial\beta} = \frac{2m_s}{\alpha_2 - \alpha_1}\delta(\alpha - \beta - w), \tag{1.57}$$

if

$$\alpha_1 \leq \alpha \leq \alpha_2, \tag{1.58}$$

and $\partial^2 f_{\alpha\beta}/\partial\alpha\partial\beta$ is equal to zero otherwise. From (1.29), (1.57), and (1.58), we find

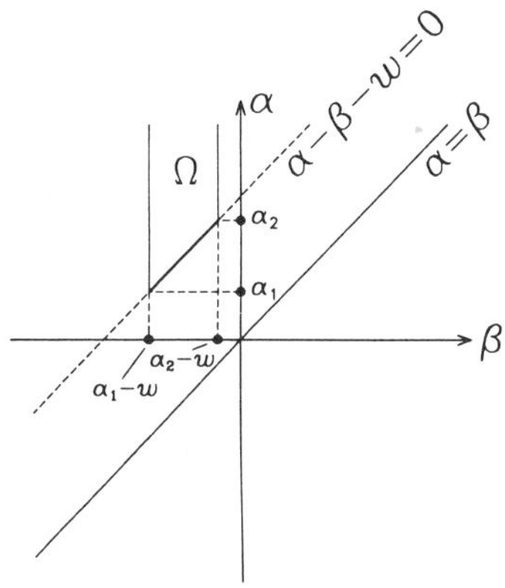

Fig. 1.33

$$\mu(\alpha, \beta) = \frac{m_s}{\alpha_2 - \alpha_1} \delta(\alpha - \beta - w) \tag{1.59}$$

for

$$\alpha_1 \leq \alpha \leq \alpha_2, \tag{1.60}$$

and $\mu(\alpha, \beta)$ is equal to zero otherwise.

Thus, the weight function $\mu(\alpha, \beta)$ has the support on the line $\alpha - \beta - w = 0$. This support is the bold segment (see Fig. 1.33) of the above line.

By substituting (1.59) and (1.60) into (1.1), after simple transformations we obtain

$$f(t) = \frac{m_s}{\alpha_2 - \alpha_1} \int_{\alpha_1}^{\alpha_2} \hat{\gamma}_{\alpha, \alpha - w} u(t) d\alpha. \tag{1.61}$$

This is the Preisach representation of the hysteresis nonlinearity shown in Fig. 1.32. It is left to the reader as a useful exercise to verify that the formula (1.61) indeed describes all branches of the above hysteresis nonlinearity.

It is clear from Fig. 1.33 that by varying α_1, α_2, and w we can freely move around the support of the function μ in the half-plane $\alpha \geq \beta$. This fact and the expression (1.61) suggest that the equivalent representation for the Preisach model can be obtained as a superposition (parallel connection) of an infinite number of hysteresis nonlinearities shown in Fig. 1.32. However, the representation (1.1) is preferable due to the most elementary nature of hysteresis operators $\hat{\gamma}_{\alpha\beta}$.

It is not accidental that in the above example the function $\mu(\alpha, \beta)$ has a "line support." This is related to the fact that the above hysteresis nonlinearity has a local memory. It can be even proved in general that if we have a hysteresis nonlinearity with a local memory which can be represented by the Preisach model, then the weight function $\mu(\alpha, \beta)$ has a support along some particular curve. This curve may intersect all possible staircase interfaces $L(t)$ only once.

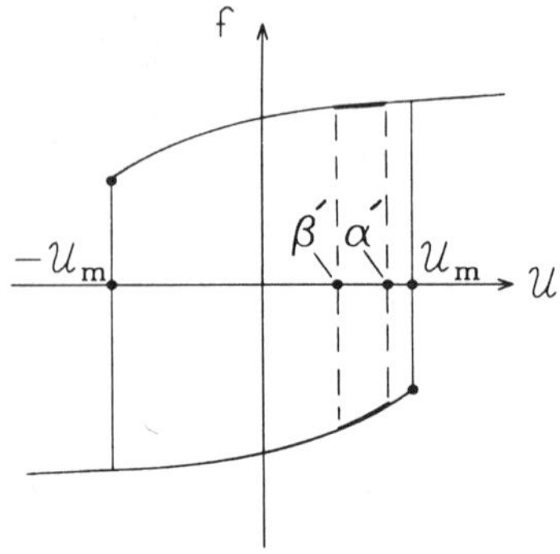

Fig. 1.34

It was emphasized before that the Preisach model can describe (and usually does describe) hysteresis with nonlocal memories. These nonlinearities are much more complex than those with local memories. However, it would be a mistake to think that any hysteresis nonlinearity with local memory can be represented by the Preisach model. In fact, there are very simple hysteresis nonlinearities with local memories which can not be described by the model. For instance, this is true for the hysteresis nonlinearity shown in Fig. 1.34. Indeed, let the input vary back and forth between two consecutive extremum values α' and β', $-u_m < \beta' < \alpha' < u_m$. Then, depending on the past history, the bold reversible parts of descending and ascending bracnches will be traced back and forth. These parts can be construed as degenerated minor loops. Since these parts are not congruent, the congruency property of the representation theorem is not satisfied. Consequently, the hysteresis nonlinearity shown in Fig. 1.34 can not be represented by the Preisach model. However, this hysteresis nonlinearity is typical for single Stoner-Wohlfarth magnetic particles. It is known from the theory of these particles that the reversible parts of their hysteresis loops result from uniform rotations of magnetization within these particles, that is, from reversible processes. Incongruency of the above parts is the reason why the Preisach model can not describe the hysteresis loops of single Stoner-Wohlfarth particles. For this reason, it can be asserted that the Preisach model does not account for reversible processes. It will be shown in the next chapter that this deficiency of the Preisach model can be removed by its appropriate generalization.

4 Numerical Implementation of the Preisach Model

The Preisach model can be numerically implemented by using the formula (1.8) for the computation of the output $f(t)$ and the formula (1.29) for the determination of the weight function $\mu(\alpha, \beta)$. Although the above approach is straightforward, it encounters two main difficulties. First, it requires the numerical evaluation of double integrals in (1.8). This is a time-consuming procedure which may impede the use of the Preisach model in practical applications. Second, the determination of the weight function $\mu(\alpha, \beta)$ by employing the formula (1.29) requires differentiations of experimentally obtained data. These differentiations may strongly amplify errors (noise) inherently present in any experimental data. It turns out that another approach can be developed for the numerical implementation of the Preisach model. This approach completely circumvents the above difficulties. It is based on the explicit formula for the integrals in (1.8). This formula directly involves (without any differentiation) the experimental data used for the identification of $\mu(\alpha, \beta)$. Moreover, the above formula will be a valuable tool for the theoretical investigation of the Preisach model, and it will be extensively used later in the discussion of vector models of hysteresis (see Chapter III).

The starting point for the derivation of the explicit formula for $f(t)$ is the expression (1.8). It is worthwhile to remind here that the positive ($S^+(t)$) and negative ($S^-(t)$) sets in (1.8) are separated by the staircase interface $L(t)$. This interface has vertices whose α and β coordinates are equal to M_k and m_k, respectively (see Fig. 1.35). As discussed in Section 2, numbers M_k and m_k form the alternating series of dominant input extrema.

By adding and subtracting the integral of $\mu(\alpha, \beta)$ over $S^+(t)$, the expression (1.8) can be represented in the form

$$f(t) = -\iint_T \mu(\alpha, \beta) d\alpha d\beta + 2 \iint_{S^+(t)} \mu(\alpha, \beta) d\alpha d\beta, \tag{1.62}$$

where, as before, T is the limiting triangle.

According to (1.26), we find

$$\iint_T \mu(\alpha, \beta) d\alpha d\beta = F(\alpha_0, \beta_0). \tag{1.63}$$

The positive set $S^+(t)$ can be subdivided into n trapezoids Q_k (see Fig. 1.35). As a result, we have

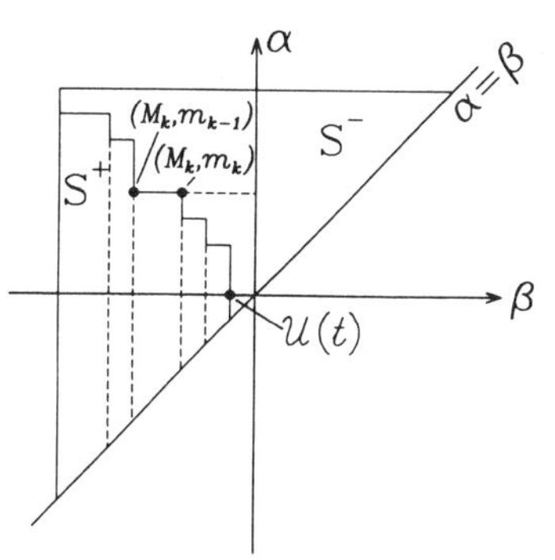

Fig. 1.35

$$\iint_{S^+(t)} \mu(\alpha, \beta)\,d\alpha\,d\beta = \sum_{k=1}^{n(t)} \iint_{Q_k(t)} \mu(\alpha, \beta)\,d\alpha\,d\beta. \tag{1.64}$$

It is clear that the number n of these trapezoids and their shapes may change with time. For this reason, n and Q_k are shown in (1.64) as functions of time.

Each trapezoid Q_k can be represented as a difference of two triangles, $T(M_k, m_{k-1})$ and $T(M_k, m_k)$. Thus, we obtain

$$\iint_{Q_k(t)} \mu(\alpha, \beta)\,d\alpha\,d\beta = \iint_{T(M_k, m_{k-1})} \mu(\alpha, \beta)\,d\alpha\,d\beta - \iint_{T(M_k, m_k)} \mu(\alpha, \beta)\,d\alpha\,d\beta, \tag{1.65}$$

where, for the case of $k = 1$, m_0 in (1.65) is naturally equal to β_0.

According to (1.26), we find

$$\iint_{T(M_k, m_{k-1})} \mu(\alpha, \beta)\,d\alpha\,d\beta = F(M_k, m_{k-1}), \tag{1.66}$$

and

$$\iint_{T(M_k, m_k)} \mu(\alpha, \beta)\,d\alpha\,d\beta = F(M_k, m_k). \tag{1.67}$$

From (1.65), (1.66), and (1.67), we derive

$$\iint_{Q_k(t)} \mu(\alpha, \beta)\,d\alpha\,d\beta = F(M_k, m_{k-1}) - F(M_k, m_k). \tag{1.68}$$

From (1.62), (1.63), (1.64), and (1.68), we obtain

$$f(t) = -F(\alpha_0, \beta_0) + 2 \sum_{k=1}^{n(t)} [F(M_k, m_{k-1}) - F(M_k, m_k)]. \tag{1.69}$$

It is clear from Fig. 1.35 that m_n is equal to the current value of input

$$m_n = u(t). \tag{1.70}$$

Consequently, the expression (1.69) can be written as

$$f(t) = -F(\alpha_0, \beta_0) + 2 \sum_{k=1}^{n(t)-1} [F(M_k, m_{k-1}) - F(M_k, m_k)]$$
$$+ 2[F(M_n, m_{n-1}) - F(M_n, u(t))]. \tag{1.71}$$

The last expression has been derived for monotonically decreasing input, that is, when the final link of interface $L(t)$ is a vertical one. If the input $u(t)$ is being monotonically increased, then the final link of $L(t)$ is a horizontal one and the $\alpha - \beta$ diagram shown in Fig. 1.35 should be slightly modified. The appropriate diagram is shown in Fig. 1.36. This diagram can be considered as a particular case of the previous one. This case is realized when

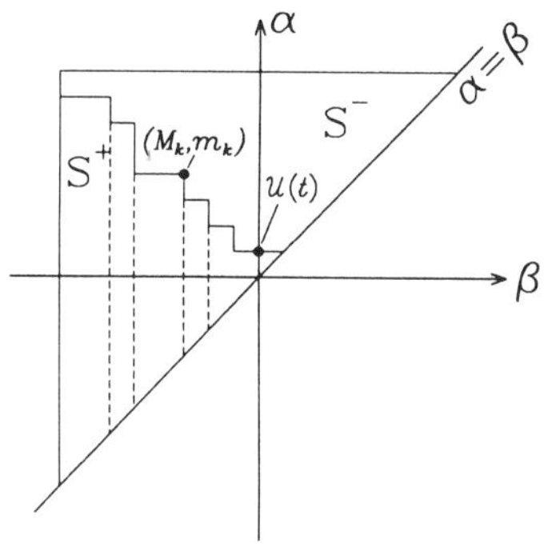

Fig. 1.36

$$m_n(t) = M_n(t) = u(t). \tag{1.72}$$

According to the definition (1.26) of $F(\alpha, \beta)$, we find

$$F(M_n, m_n) = F(u(t), u(t)) = 0. \tag{1.73}$$

From (1.69), (1.72), and (1.73), we derive the following expression for $f(t)$ in the case of monotonically increasing input:

$$f(t) = -F(\alpha_0, \beta_0) + 2\sum_{k=1}^{n(t)-1} [F(M_k, m_{k-1}) - F(M_k, m_k)] + 2F(u(t), m_{n-1}). \tag{1.74}$$

The function $F(\alpha, \beta)$ is related to experimentally measured first-order transition curves by the formula (1.24). Using this formula, expressions (1.71) and (1.74) can be written in terms of the above experimental data as follows:

$$f(t) = -f^+ + \sum_{k=1}^{n-1} (f_{M_k m_k} - f_{M_k m_{k-1}}) + f_{M_n u(t)} - f_{M_n m_{n-1}}, \tag{1.75}$$

$$f(t) = -f^+ + \sum_{k=1}^{n-1} (f_{M_k m_k} - f_{M_k m_{k-1}}) + f_{-m_{n-1}} - f_{-m_{n-1}, -u(t)}. \tag{1.76}$$

Here, f^+ is, as before, the positive saturation value of output, and the last term in (1.74) has been transformed by using the formulas (1.24) and (1.44). Thus, we have derived the explicit expressions (1.75) and (1.76) for $f(t)$ in terms of experimentally measured data. These expressions constitute the basis for the numerical implementation of the Preisach model.

Using the above expressions, the digital code which implements the Preisach model has been developed. This code determines the output of a hysteresis transducer from an input, a set of first-order transition curves and an input history which are all specified by the user. The code also allows the user the flexibility of specifying an analytical expression for the weight function $\mu(\alpha, \beta)$ instead of first-order transition curves. In this case, the function $F(\alpha, \beta)$ defined by (1.26) is precomputed and then used in the formulas (1.71) and (1.74) for the evaluation of output values.

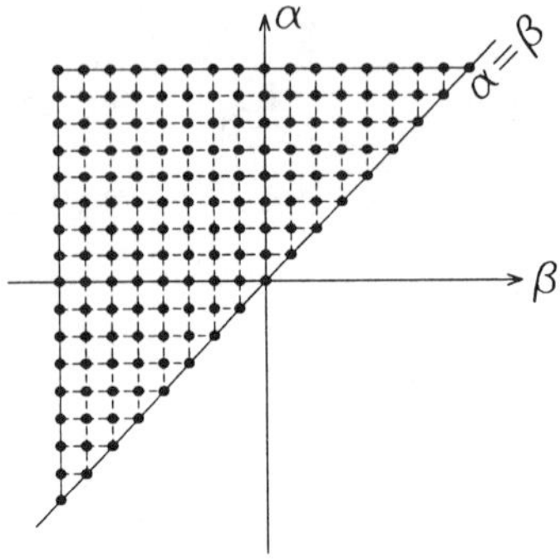

Fig. 1.37

As usual, the code performs three main functions: preprocessing, processing, and postprocessing. During the preprocessing state, a square mesh covering the limiting triangle T (see Fig. 1.37) is created and a discretized set of first-order transition curves is entered. This set consists of mesh values of the function $f_{\alpha\beta}$. In the case when an analytical expression for $\mu(\alpha, \beta)$ is specified, the mesh values of $F(\alpha, \beta)$ are computed during the preprocessing. At the processing stage, an input history and current values of input are entered. Using these data and the formulas (1.19) and (1.20), the alternating series of dominant input extrema $\{M_k, m_k\}$ is first determined and then continuously updated for each new instant of time. Using M_k, m_k and mesh values of $f_{\alpha\beta}$ (or $F(\alpha, \beta)$), all terms in the formulas (1.75) and (1.76) (or (1.71) and (1.74)) are computed. This is done by first determining particular square (or triangular) cells to which points (M_k, m_k), (M_k, m_{k-1}), and $(M_n, u(t))$ belong, and then by computing the values of $f_{\alpha\beta}$ (or $F(\alpha, \beta)$) at these points by means of interpolation of the mesh values of $f_{\alpha\beta}$ (or $F(\alpha, \beta)$) at the vertices of the above cells. In the code, the interpolation is based on the following local approximation to $f_{\alpha\beta}$ in each square mesh cell:

$$f_{\alpha\beta} \simeq C_0 + C_1\alpha + C_2\beta + C_3\alpha\beta. \tag{1.77}$$

The C-coefficients in (1.77) are found by matching the values of $f_{\alpha\beta}$ in the cell vertices. For triangular cells which are adjacent to the line $\alpha = \beta$, the linear local approximations to $f_{\alpha\beta}$ are used:

$$f_{\alpha\beta} \simeq C_0 + C_1\alpha + C_2\beta. \tag{1.78}$$

After computing all necessary values of $f_{\alpha\beta}$, the current values of output are evaluated by employing the formulas (1.75) and (1.76). At the postprocessing stage, the computed output values are stored in a special file along with the corresponding input values and the values of M_k and m_k. The user has the option to plot the output and input as functions of time, the path traced on the $f - u$ plane and the set of first-order transition curves used in computations.

The code described above is written in the structured language, FORTRAN 77, and it utilizes the Tektronix PLOT-10 graphic package. It has been

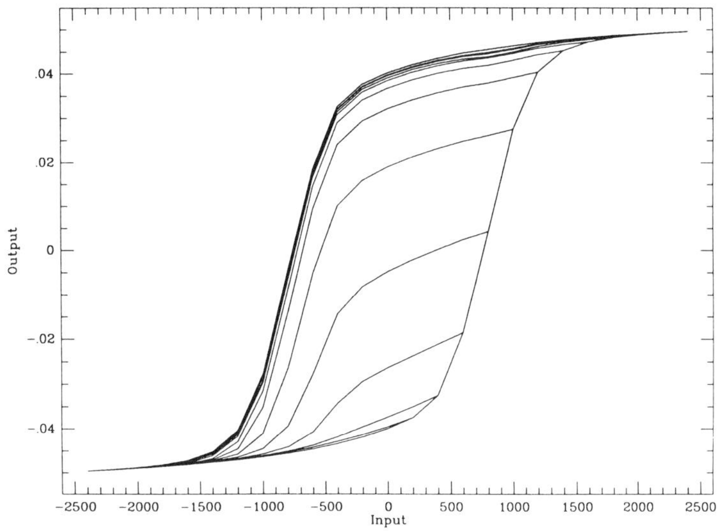

Fig. 1.38

developed on VAX 11/780 operating under the UNIX 4.2 BSD system. This code will compile and run on any computer which supports FORTRAN 77 and PLOT-10. Below, we shall present some examples computed by using the above code. Figures 1.38, 1.39, and 1.40 show the set of first-order transition curves, the initial staircase interface $L(t)$, and the input $u(t)$, respectively. These data have been used to compute the output $f(t)$. The computed

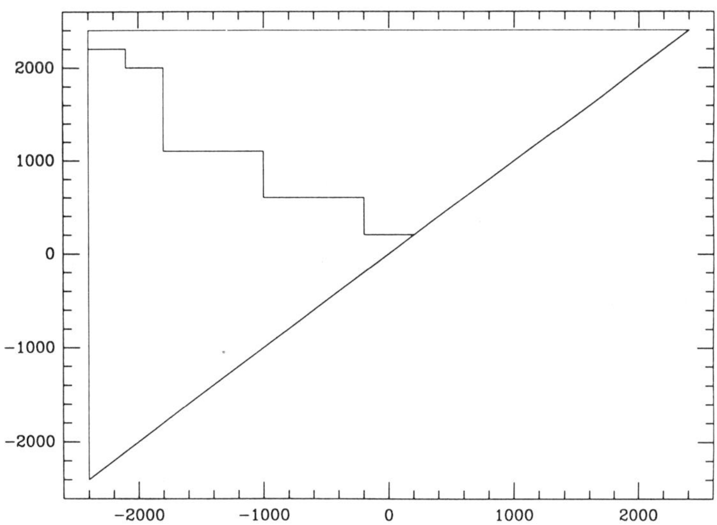

Fig. 1.39

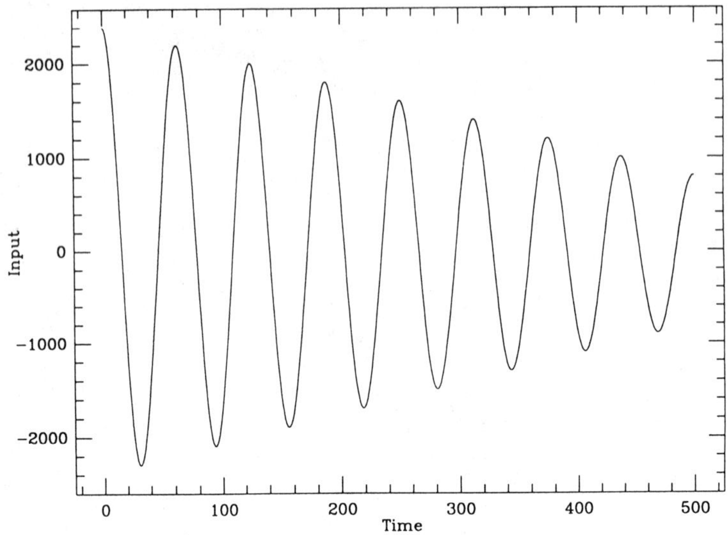

Fig. 1.40

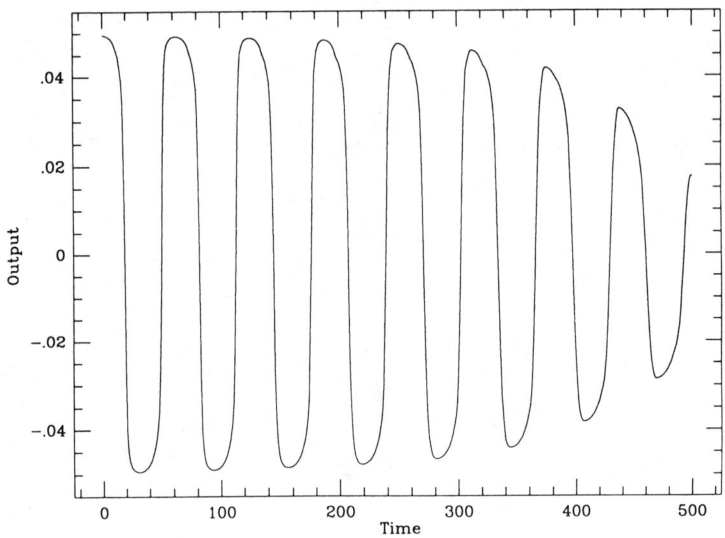

Fig. 1.41

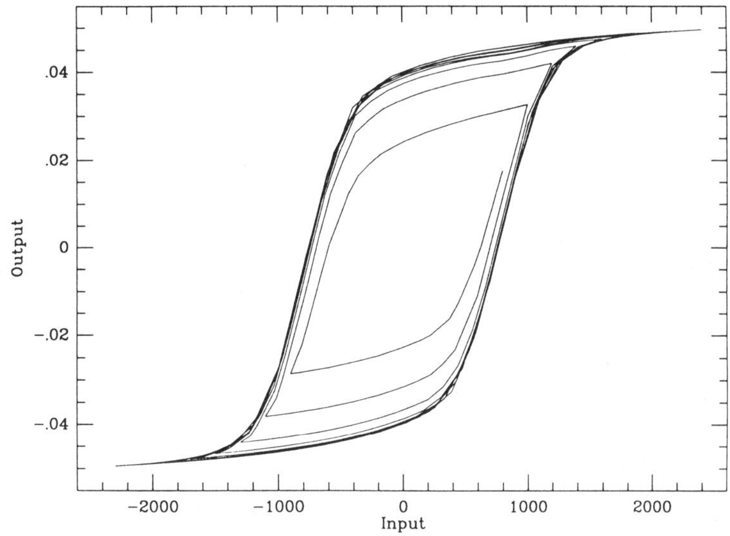

Fig. 1.42

output $f(t)$ and the $f - u$ path of the transducer are given in Fig. 1.41 and 1.42, respectively.

As was pointed out before, the derived formulas (1.75) and (1.76) (or (1.71) and (1.74)) are useful not only for the numerical implementation of the Preisach model but for its theoretical investigation as well. For instance, by using (1.75) and (1.76), the following proposition can be easily proven.

Theorem

Any $f - u$ path of hysteresis transducer representable by the Preisach model is piecewise congruent to first-order transition curves. In other words, any $f - u$ path of such a transducer consists of pieces, which are congruent to some particular first-order transition curves.

Proof

Suppose that the input $u(t)$ is being monotonically decreased from its previous maximum value $u = M_n$ until it reaches the value $u = m_{n-1}$. Then, the formula (1.75) is valid. In this formula, all terms except $f_{M_n u(t)}$ do not vary. Consequently, the $f - u$ path traced during the above input variation is congruent to the first-order transition curve $f_{M_n u}$ which is attached to the limiting ascending branch at the point f_{M_n}. As the input reaches the value m_{n-1}, the cancellation of the last two terms in (1.75) occurs. If the input is being further decreased remaining between $u = m_{n-1}$ and $u = m_{n-2}$, then the formula (1.75) is modified as follows.

$$f(t) = -f_+ + \sum_{k-1}^{n-2} (f_{M_k m_k} - f_{M_k m_{k-1}}) + f_{M_{n-1} u(t)} - f_{M_{n-1} m_{n-2}}. \qquad (1.79)$$

From this formula, as before, we conclude that the $f - u$ path traced during the above input variation is congruent to the first-order transition curve $f_{M_{n-1} u}$ which is attached to the limiting ascending branch at the point $f = f_{M_{n-1}}$. By continuing the same line of reasoning, we find that if the input is being monotonically decreased between $u = m_k$ and $u = m_{k-1}$, then the corresponding $f - u$ path is congruent to the first-order transition curve $f_{M_k u}$. For the case when the input is monotonically increased, the formula (1.76) is valid. From this formula, we find that if the input is monotonically increased between the values $u = M_k$ and $u = M_{k-1}$, then the time varying part of the output is described by the term $f_{-m_{k-1}, -u(t)}$. Thus, the corresponding $f - u$ path is congruent to the first-order transition curve $f_{-m_{k-1}, -u}$ which is attached to the limiting ascending branch at the point $f = f_{-m_{k-1}}$. This completes the proof.

It is evident from the proven theorem that initial slopes of reversal curves described by the Preisach model coincide with initial slopes of experimentally measured first-order transition curves, and, consequently, in general they are not equal to zero. This is consistent with our discussion of initial slopes of reversal curves presented in the previous section.

Another important application of the formulas (1.75) and (1.76) is the derivation of the expressions for linear extrapolations of output corresponding to small input increments. Such extrapolative expressions can be useful for numerical analysis of systems with hysteresis by time-marching techniques. The derivation of these expressions proceeds as follows. Consider the case when the input has been monotonically increased for $t' \leq t$ and this monotonic increase continues just after the time t. Then, the formula (1.76) is valid for the output before and after the time t. From this formula, by using the local Taylor expansion at time t for $t' > t$, we find

$$f(t') - f(t) \simeq -\left.\frac{\partial f_{-m_{n-1}\beta}}{\partial \beta}\right|_{\beta = -u(t)} \cdot [u(t') - u(t)]. \qquad (1.80)$$

It is convenient to represent the last formula in terms of input and output increments:

$$\Delta f = -\left.\frac{\partial f_{-m_{n-1}\beta}}{\partial \beta}\right|_{\beta = -u(t)} \cdot \Delta u. \qquad (1.81)$$

It is important to keep in mind that m_{n-1} in (1.81) may change with time. This occurs because of the wiping-out property. It is easy to see that, at any instant of time, m_{n-1} in (1.81) is the largest m_k in the series of M_k and m_k.

If a monotonic increase of input for $t' \leq t$ is followed by a monotonic input decrease for $t' > t$, then the input value $u(t)$ can be treated as the last maximum,

M_n, and from (1.75), we find

$$f(t) = -f_+ + \sum_{k=1}^{n-1} (f_{M_k m_k} - f_{M_k m_{k+1}}) + f_{u(t)u(t)} - f_{u(t)m_{n-1}} \tag{1.82}$$

and

$$f(t') = -f_+ + \sum_{k=1}^{n-1} (f_{M_k m_k} - f_{M_k m_{k-1}}) + f_{u(t)u(t')} - f_{u(t)m_{n-1}}. \tag{1.83}$$

From (1.82) and (1.83), we obtain

$$f(t') - f(t) = f_{u(t)u(t')} - f_{u(t)u(t)}. \tag{1.84}$$

Using the last expression, we derive

$$\Delta f \simeq \left. \frac{\partial f_{u(t)\beta}}{\partial \beta} \right|_{\beta = u(t)} \cdot \Delta u. \tag{1.85}$$

Now, we consider the case when the input has been undergoing a monontonic decrease for $t' \leq t$. If this decrease continues for $t' > t$, then the formula (1.75) is valid. From this formula as before we find

$$\Delta f \simeq \left. \frac{\partial f_{M_n \beta}}{\partial \beta} \right|_{\beta = u(t)} \cdot \Delta u. \tag{1.86}$$

In the last formula, M_n may change with time. It is clear that at any instant of time M_n in (1.81) is the smallest M_k in the alternating series of M_k and m_k.

If the monotonic decrease of input for $t' \leq t$ is followed by a subsequent monotonic input increase for $t' > t$, then the input value $u(t)$ can be considered as m_{n-1} and the formula (1.76) can be used for $t' > t$. From this formula, as before we derive

$$\Delta f \simeq - \left. \frac{\partial f_{-u(t)\beta}}{\partial \beta} \right|_{\beta = u(t)} \cdot \Delta u. \tag{1.87}$$

Thus, for the previously increasing input, the output can be linearly extrapolated by using formulas (1.81) and (1.85) for $\Delta u > 0$ and $\Delta u < 0$, respectively. For the previously decreasing input, the output can be linearly extrapolated by using the formulas (1.86) and (1.87) for $\Delta u < 0$ and $\Delta u > 0$, respectively. It is important to note that the above formulas are given in terms of slopes of experimentally measured first-order transition curves.

The formulas (1.81), (1.85), (1.86) and (1.87) give *local* linearizations of output increments at time t. In other words, these linearizations are fairly accurate only for small time difference $t - t'$. However, by employing the formulas (1.75) and (1.76), it is possible to find global linearizations of output increments which are valid for finite (not necessary small) time intervals. The simplest way to do this is to use a special set of continuous inputs $u(t)$. Consider an arbitrary time interval $[t_0, t']$. We define the set $K_{[t_0, t']}$ of inputs which are continuous on this interval and have the properties: (a) any two inputs from the set assume their maximum and minimum values at the same instances of time, (b) maxima and minima which constitute alternating series of dominant extrema are also assumed at the same times. For two inputs $u(t)$ and $\tilde{u}(t)$, the last condition can be expressed as

$$u(t_k^+) = M_k, \qquad \tilde{u}(t_k^+) = \tilde{M}_k, \tag{1.88}$$

$$u(t_k^-) = m_k, \qquad \tilde{u}(t_k^-) = \tilde{m}_k, \tag{1.89}$$

where the same notations as in Section 2 are used above.

It is obvious that $K_{[t_0,t']}$ is a convex set of functions. Indeed, if the above two properties are satisfied for inputs $u(t)$ and $\tilde{u}(t)$, they are also satisfied for the input $\lambda u(t) + (1 - \lambda)\tilde{u}(t)$ where $0 \le \lambda \le 1$. It is also clear that if $u(t)$ belongs to $K_{[t_0,t']}$, then $\lambda u(t)$ also belongs to $K_{[t_0,t']}$ for any $\lambda \ge 0$ and it does not belong to $K_{[t_0,t']}$ for $\lambda < 0$. This means that the set $K_{[t_0,t']}$ is a cone in the Banach space of continuous functions. It is clear that the set of inputs $K_{[t_0,t']}$ depends on the alternating sequence of t_k^+ and t_k^-. By specifying different times t_k^+ and t_k^-, we end up with different sets of inputs. For this reason, it would be more precise to adopt the notation $K_{[t_0,t_1^+,t_1^-,\ldots,t_n^+,t_n^-,t']}$. But for the sake of conciseness, we preserve the previous notation.

From the above definition of $K_{[t_0,t']}$, it follows that any two inputs from $K_{[t_0,t']}$ are monotonically decreased (or increased) on the same time intervals. This means that at any instant of time the same formula (1.75) (or (1.76)) can be applied to the computations of outputs. Thus, assuming that at time t both inputs $u(t)$ and $\tilde{u}(t)$ are monotonically decreased, we can use (1.75) for $u(t)$ and the similar expression for $\tilde{u}(t)$:

$$\tilde{f}(t) = -f^+ + \sum_{k=1}^{n-1} (f_{\tilde{M}_k\tilde{m}_k} - f_{\tilde{M}_k\tilde{m}_{k-1}}) + f_{\tilde{M}_n\tilde{u}(t)} - f_{\tilde{M}_n\tilde{m}_n}. \tag{1.90}$$

By subtracting (1.75) from (1.90) and retaining only linear terms of Taylor expansions, we derive

$$\begin{aligned}\delta f(t) = &\sum_{k=1}^{n-1}\left(\frac{\partial f_{\alpha\beta}}{\partial\alpha}\bigg|_{M_k,m_k}\cdot\delta M_k + \frac{\partial f_{\alpha\beta}}{\partial\beta}\bigg|_{M_k,m_k}\cdot\delta m_k\right)\\ &-\sum_{k=1}^{n}\left(\frac{\partial f_{\alpha\beta}}{\partial\alpha}\bigg|_{M_k,m_{k-1}}\cdot\delta M_k + \frac{\partial f_{\alpha\beta}}{\partial\beta}\bigg|_{M_k,m_{k-1}}\cdot\delta m_{k-1}\right)\\ &+\frac{\partial f_{\alpha\beta}}{\partial\alpha}\bigg|_{M_n,u(t)}\cdot\delta M_n + \frac{\partial f_{\alpha\beta}}{\partial\beta}\bigg|_{M_n,u(t)}\cdot\delta u(t),\end{aligned} \tag{1.91}$$

where the following notations are used:

$$\delta f(t) = \tilde{f}(t) - f(t), \qquad \delta u(t) = \tilde{u}(t) - u(t), \tag{1.92}$$

$$\delta M_k = \tilde{M}_k - M_k, \qquad \delta m_k = \tilde{m}_k - m_k. \tag{1.93}$$

If both inputs are monotonically increased at time t, then the formula (1.76) is applied and, as before, we derive

$$
\begin{aligned}
\delta f(t) = \sum_{k=1}^{n-1} \Bigg(& \frac{\partial f_{\alpha\beta}}{\partial \alpha}\bigg|_{M_k, m_k} \cdot \delta M_k + \frac{\partial f_{\alpha\beta}}{\partial \beta}\bigg|_{M_k, m_k} \cdot \delta m_k \\
& - \frac{\partial f_{\alpha\beta}}{\partial \alpha}\bigg|_{M_k, m_{k-1}} \cdot \delta M_k - \frac{\partial f_{\alpha\beta}}{\partial \beta}\bigg|_{M_k, m_{k-1}} \cdot \delta m_{k-1} \Bigg) \\
& - \left(\frac{\partial f_{\alpha}}{\partial \alpha}\bigg|_{-m_{n-1}} - \frac{\partial f_{\alpha\beta}}{\partial \alpha}\bigg|_{-m_{n-1}, -u(t)} \right) \delta m_{n-1} + \frac{\partial f_{\alpha\beta}}{\partial \beta}\bigg|_{-m_{n-1}, -u(t)} \cdot \delta u(t). \qquad (1.94)
\end{aligned}
$$

The expressions (1.91) and (1.94) represent a global linearization of output increment on the interval $[t_0, t']$. These expressions can be construed as the Frechet derivative of the Preisach hysteresis operator on the cone $K_{[t_0, t']}$. These expressions may be useful in the analysis of periodic regimes in systems with hysteresis by perturbation techniques.

We conclude this section by discussing another example which demonstrates the usefulness of the above explicit formulas for the output $f(t)$. Consider an input $u(t)$ which has an infinite number of oscillations as time progresses to infinity. We assume that the amplitude of these oscillations (around some value $u(\infty)$) monotonically diminishes to zero. In other terms, this means that

$$\lim_{k \to \infty} M_k = \lim m_k = u(\infty) = \xi. \qquad (1.95)$$

The described input variations are typical for anhysteretic magnetization processes.

We are going to prove that there exists the limit for the output $f(t)$ as time goes to infinity.

By using (1.75), we find

$$\lim_{t \to \infty} f(t) = -F(\alpha_0, \beta_0) + 2 \sum_{k=1}^{\infty} [F(M_k, m_{k-1}) - F(M_k, m_k)]. \qquad (1.96)$$

Thus, it remains to be proven that the infinite series in (1.96) converges. To this end, we introduce the following notations:

$$F(M_k, m_{k-1}) = a_{2k}, \qquad F(M_k, m_k) = a_{2k+1} \qquad (1.97)$$

and rewrite (1.96) as

$$\lim_{t \to \infty} f(t) = -F(\alpha_0, \beta_0) + 2 \sum_{k=2}^{\infty} (-1)^k a_k. \qquad (1.98)$$

Since usually $\mu(\alpha, \beta) \geq 0$, from (1.26) and (1.97) we find that $a_k > 0$. Moreover, since

$$m_{k-1} < m_k \quad \text{and} \quad M_k > M_{k+1}, \qquad (1.99)$$

we conclude that

$$a_{2k} = F(M_k, m_{k-1}) > F(M_k, m_k) = a_{2k+1}, \tag{1.100}$$

$$a_{2k+1} = F(M_k, m_k) > F(M_{k+1}, m_k) = a_{2k+2}. \tag{1.101}$$

In addition, from (1.95) and (1.26) we have

$$\lim_{k\to\infty} a_{2k} = \lim_{k\to\infty} a_{2k+1} = F(\xi, \xi) = 0.$$

Thus, the expression (1.98) is the alternating series with monotonically decreasing to zero coefficients. According to the well-known theorem, this series converges. This proves that there exists the limit of $f(t)$ for $t \to \infty$.

5 The Preisach Model and Hysteretic Energy Losses

A hysteresis phenomenon is associated with some energy dissipation which in magnetics is often referred to as hysteretic energy losses. The problem of determining hysteretic energy losses is a classical one. It has been attracting considerable attention because the hysteresis energy loss is an important component of core losses occurring in almost all electromagnetic power devices as well as in many high frequency microwave devices. For this reason, the means for accurate predictions of hysteresis losses and their reduction are important for the optimal design of various equipment. The solution to the above problem has long been known for the particular case of periodic (cyclic) input variations. In magnetics, this solution is most often associated with the name of C.P. Steinmetz. This solution shows that a hysteretic energy loss is equal to an area enclosed by a loop resulting from periodic input variations (see Fig. 1.43). However, energy dissipation is a continuous process, and it occurs for arbitrary (not necessary periodic) variations of input. The problem of computing hysteretic energy losses for the above general case has remained unsolved. A solution to this problem would be of both theoretical and practical importance. From the theoretical point of view, the solution to the above problem will allow for the calculation of internal entropy production which is a key point in the development of irreversible thermodynamics of hysteretic

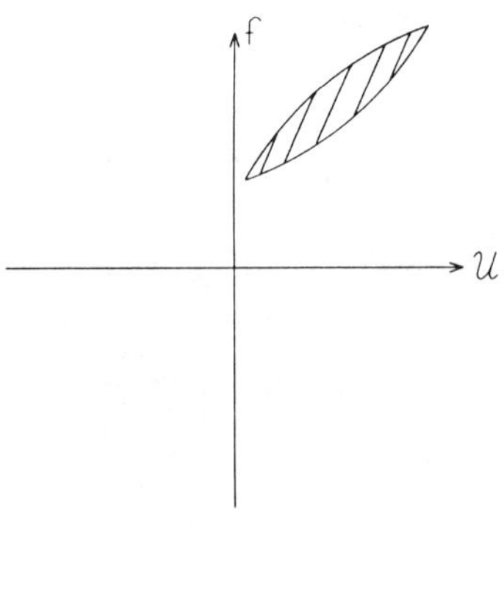

Fig. 1.43

media. In magnetics, it may also allow one to separate a dissipated energy from an energy stored in magnetic field. This eventually may lead to expressions for electromagnetic forces in hysteretic media. From the practical viewpoint, the solution to the above problem may bring new experimental techniques for the measurement of hysteretic energy losses occurring for arbitrary input variations.

It should not be surprising that the expression for hysteretic energy losses has been found only for the case of periodic input variations. The reason behind this fact is that the hysteretic energy losses occurring for periodic input variations can be easily evaluated by using only the energy conservation principle; no knowledge of actual mechanisms of hysteresis or its model is required. The situation is much more complicated when arbitrary input variations are considered. Here, the energy conservation principle alone is not sufficient, and an adequate model of hysteresis should be employed in order to arrive at the solution to the problem. It turns out that the Preisach hysteresis model is very well suited for this purpose.

In this section, the Preisach model will be used for the derivation of general expressions for hysteretic energy losses. These expressions will be given in terms of the weight function $\mu(\alpha, \beta)$ as well as in terms of experimentally measured first-order transition curves. Furthermore, a formula which relates the hysteretic energy losses occurring for arbitrary input variations to the losses occurring for certain periodic input variations will be found. This formula may result in simple techniques for the measurement of hysteretic losses occurring for arbitrary input variations. The application of the mentioned results to the irreversible thermodynamics of hysteretic media will also be discussed.

We begin by defining the input $u(t)$ and the output $f(t)$ as work variables. This means that the infinitesimal energy supplied to the transducer (media) in the form of work is given by

$$\delta W = u df. \tag{1.102}$$

In magnetics, u is the magnetic field H, f is the magnetization M, and the formula (1.102) becomes the classical expression for the work done in magnetizing a unit volume of magnetic media:

$$\delta W = H dM. \tag{1.103}$$

Similarly, in mechanics, u is the force F, f is the specific length L, and from (1.102) we find the standard formula

$$\delta W = F dL. \tag{1.104}$$

Now, we proceed to the derivation of expressions for hysteretic energy losses. We first consider the case when a hysteresis nonlinearity is represented by a rectangular loop shown in Fig. 1.44. If a periodic variation of input is such that the whole loop is traced, then the hysteretic energy loss for one cycle, Q_{cycle}, is equal to the area enclosed by the loop

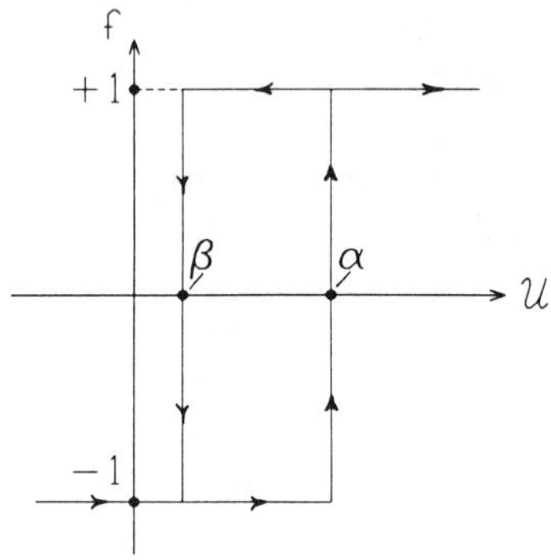

Fig. 1.44

$$Q_{\text{cycle}} = 2(\alpha - \beta). \tag{1.105}$$

It is clear that the horizontal links of the loop are fully reversible, and, for this reason, no energy losses occur as these links are traced. Thus, it can be concluded that only "switching-up" and "switching-down" result in energy losses. It can be assumed (on the physical grounds) that there is symmetry between the above two switchings. In other words, these switchings are identical as far as energy losses are concerned. Consequently, the same energy loss occurs for each of these switchings. As a result, we conclude that the energy loss per switching q is given by

$$q = \alpha - \beta. \tag{1.106}$$

The product $\mu(\alpha, \beta)\hat{\gamma}_{\alpha\beta}$ can be construed as a rectangular hysteresis loop with output values equal to $\pm\mu(\alpha, \beta)$. For this reason, switchings of such loops will result in energy losses equal to $\mu(\alpha, \beta)(\alpha - \beta)$. In the Preisach model, any input variation is associated with switchings of some rectangular loops $\mu(\alpha, \beta)\hat{\gamma}_{\alpha\beta}$. These switchings represent irreversible processes occurring during input variations. Consequently, it is natural to equate the hysteretic energy loss occurring for some input variation to the sum of energy losses resulting from the switching of rectangular loops during this input variation. Since in the Preisach model we are dealing with continuous ensembles of rectangular loops, the above summation should be replaced by integration. Thus, if Ω denotes the region of points on $\alpha - \beta$ diagram for which rectangular loops were switched during some input variation, then the hysteretic energy loss Q for this input variation is given by

$$Q = \iint_{\Omega} \mu(\alpha, \beta)(\alpha - \beta)\,d\alpha\,d\beta. \tag{1.107}$$

This is the fundamental formula for hysteretic energy losses, and all subsequent results will follow from this expression. It is clear from the above reasoning that the derivation of the formula (1.107) rests on the following two facts: (a) for rectangular hysteresis loops hysteretic losses can be evaluated for arbitrary

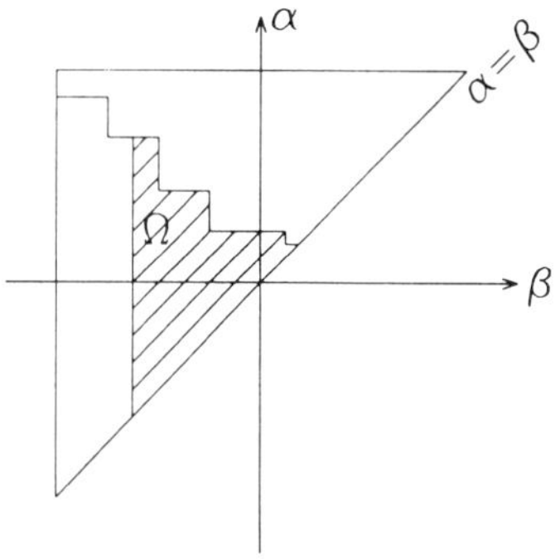

Fig. 1.45

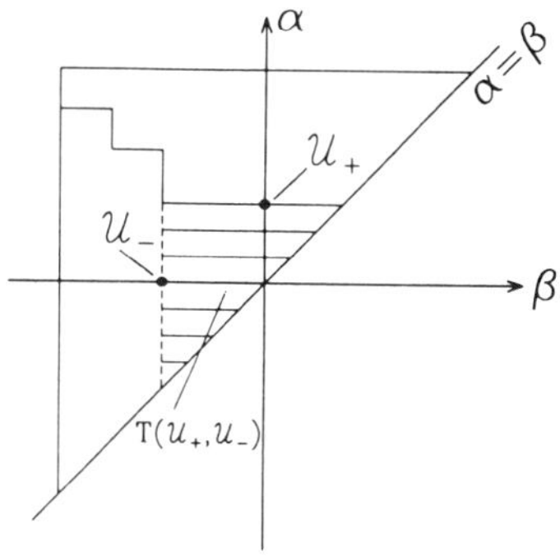

Fig. 1.46

input variations, and (b) the Preisach model represents complicated hysteresis nonlinearities as superpositions of rectangular loops. The above two facts make the Preisach model a very convenient tool for the solution of the problem at hand. However, it has to be kept in mind that the formula (1.107) can not be applied to any hysteresis nonlinearity. It has certain limits of applicability which are the same as for the Preisach model itself.

A typical shape of the region Ω is shown in Fig. 1.45. It is clear from this figure that Ω can be always subdivided into a triangle and some trapezoids. The trapezoids, in turn, can be represented as differences of triangles. Thus, if the integral in (1.107) can be evaluated for any triangular region, then it will be easy to determine this integral for any possible shape of Ω. For this reason, it makes sense to compute the values of the above integrals over various triangles. By using these values, hysteresis losses can be easily found for any input variations. In the case when Ω is a triangle, the integral (1.107) can be evaluated in terms of first-order transition curves. The derivation proceeds as follows.

Consider the function $(\alpha - \beta)F(\alpha, \beta)$. By differentiating this function, we find

$$\frac{\partial^2}{\partial\alpha\partial\beta}[(\alpha - \beta)F(\alpha, \beta)] = \frac{\partial F(\alpha, \beta)}{\partial\beta} - \frac{\partial F(\alpha, \beta)}{\partial\alpha} + (\alpha - \beta)\frac{\partial^2 F(\alpha, \beta)}{\partial\alpha\partial\beta}. \quad (1.108)$$

By using (1.28) and (1.108), we derive

$$\mu(\alpha, \beta)(\alpha - \beta) = \frac{\partial F(\alpha, \beta)}{\partial\beta} - \frac{\partial F(\alpha, \beta)}{\partial\alpha} - \frac{\partial^2}{\partial\alpha\partial\beta}[(\alpha - \beta)F(\alpha, \beta)]. \quad (1.109)$$

Let $T(u_+, u_-)$ be a triangle (see Fig. 1.46) swept during the monotonic increase of input from u_- to u_+. According to (1.107), the above input variation results in the hysteretic loss $Q(u_-, u_+)$ which is given by

$$Q(u_-, u_+) = \iint_{T(u_+, u_-)} \mu(\alpha, \beta)(\alpha - \beta)\,d\alpha\,d\beta. \quad (1.110)$$

Substituting the expression (1.109) into (1.110), we find

$$Q(u_-, u_+) = \iint_{T(u_+, u_-)} \frac{\partial F(\alpha, \beta)}{\partial \beta} d\alpha d\beta - \iint_{T(u_+, u_-)} \frac{\partial F(\alpha, \beta)}{\partial \alpha} d\alpha d\beta$$

$$- \iint_{T(u_+, u_-)} \frac{\partial^2}{\partial \alpha \partial \beta} [(\alpha - \beta) F(\alpha, \beta)] d\alpha d\beta. \qquad (1.111)$$

The first integral in (1.111) can be evaluated as follows:

$$\iint_{T(u_+, u_-)} \frac{\partial F(\alpha, \beta)}{\partial \beta} d\alpha d\beta = \int_{u_-}^{u_+} \left(\int_{u_-}^{\alpha} \frac{\partial F(\alpha, \beta)}{\partial \beta} d\beta \right) d\alpha$$

$$= \int_{u_-}^{u_+} F(\alpha, \alpha) d\alpha - \int_{u_-}^{u_+} F(\alpha, u_-) d\alpha$$

$$= - \int_{u_-}^{u_+} F(\alpha, u_-) d\alpha, \qquad (1.112)$$

since $F(\alpha, \alpha) = 0$.

Similarly, for the second integral in (1.111) we obtain

$$\iint_{T(u_+, u_-)} \frac{\partial F(\alpha, \beta)}{\partial \alpha} d\alpha d\beta = \int_{u_-}^{u_+} F(u_+, \beta) d\beta. \qquad (1.113)$$

Finally, for the third integral in (1.111) we derive

$$\iint_{T(u_+, u_-)} \frac{\partial^2}{\partial \alpha \partial \beta} [(\alpha - \beta) F(\alpha, \beta)] d\alpha d\beta$$

$$= \int_{u_-}^{u_+} \left(\int_{\beta}^{u_+} \frac{\partial^2}{\partial \alpha \partial \beta} [(\alpha - \beta) F(\alpha, \beta)] d\alpha \right) d\beta$$

$$= \int_{u_-}^{u_+} \frac{\partial}{\partial \beta} [(u_+ - \beta) F(u_+, \beta)] d\beta = -(u_+ - u_-) F(u_+, u_-). \qquad (1.114)$$

Substituting (1.112), (1.113) and (1.114) into (1.111), we find

$$Q(u_-, u_+) = (u_+ - u_-) F(u_+, u_-) - \int_{u_-}^{u_+} F(u_+, \beta) d\beta - \int_{u_-}^{u_+} F(\alpha, u_-) d\alpha. \qquad (1.115)$$

The last formula has two main advantages over (1.110). First, its application requires the evaluation of one-dimensional integrals. Second, this formula expresses the losses directly in terms of experimentally measured first-order transition curves.

By using the expression (1.115), the hysteretic energy losses can be evaluated for arbitrary input variations. Consider some input variation for which the region Ω has the shape shown in Fig. 1.45. Then, the corresponding energy loss is given by (1.107) which can be written as follows:

$$Q = Q(M_n, m_n) + \sum_{i=1}^{k} [Q(M_{n-i}, m_{n-i}) - Q(M_{n-i}, m_{n-i-1})], \tag{1.116}$$

where it is assumed that

$$m_{n-k-1} = u(t) \tag{1.117}$$

and each term in (1.116) can be evalated by employing the formula (1.115).

Next, we discuss some interesting qualitative properties of energy losses occurring in hysteresis transducers described by the Preisach model. Consider a cyclic variation of input between two consecutive extremum values u_- and u_+. During the monotonic increase of input from u_- to u_+, the final horizontal link of $L(t)$ sweeps the triangle $T(u_+, u_-)$ (see Fig. 1.46). Consequently, the losses occurring during this monotonic increase are given by (1.110). On the other hand, during the monotonic decrease of input from u_+ to u_-, the final vertical link of $L(t)$ sweeps the same triangle, and the corresponding losses $Q(u_+, u_-)$ will be given by the same integral in (1.110). Thus,

$$Q(u_-, u_+) = Q(u_+, u_-), \tag{1.118}$$

and we obtain the following result.

Theorem

For anly loop, the hysteretic losses occurring along ascending and descending branches are the same.

The above theorem can be used to find the formula which relates a hysteretic loss occurring for arbitrary input variations to certain cyclic hysteretic losses. Suppose that the input $u(t)$ is monotonically increased from some minimum value u_- and it reaches successively the values u_1 and u_2 with $u_2 > u_1$ (see Fig. 1.47). We are concerned with the hysteretic loss $Q(u_1, u_2)$ during the monotonic input increase between u_1 and u_2. For this input increase we have (see Fig. 1.48)

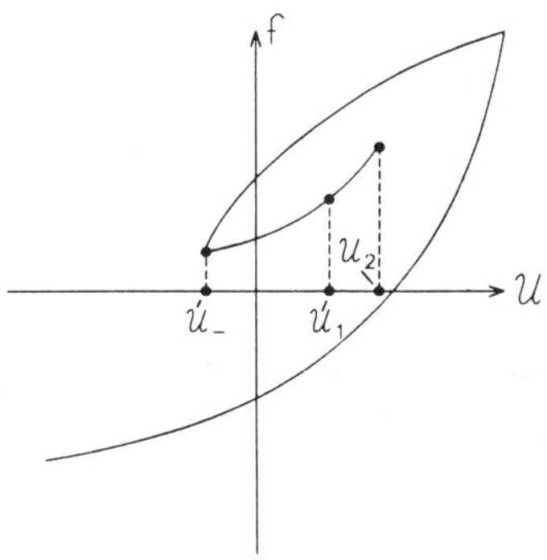

Fig. 1.47

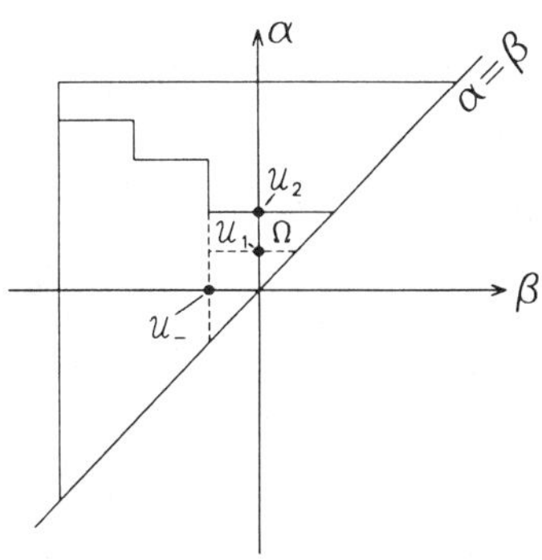

Fig. 1.48

$$\Omega = T(u_2, u_-) - T(u_1, u_-). \tag{1.119}$$

Consequently,

$$Q(u_1, u_2) = Q(u_-, u_2) - Q(u_-, u_1). \tag{1.120}$$

Using the above theorem, losses $Q(u_-, u_2)$ and $Q(u_-, u_1)$ can be expressed in terms of cyclic losses

$$Q(u_-, u_1) = \tfrac{1}{2}\bar{Q}(u_-, u_1), \qquad Q(u_-, u_2) = \tfrac{1}{2}\bar{Q}(u_-, u_2), \tag{1.121}$$

where $\bar{Q}(u_-, u_1)$ is the hysteretic loss per cycle when the input is periodically varied between u_- and u_1; the notation $\bar{Q}(u_-, u_2)$ has a similar meaning.

By substituting (1.121) into (1.120), we obtain

$$Q(u_1, u_2) = \tfrac{1}{2}[\bar{Q}(u_-, u_2) - \bar{Q}(u_-, u_1)]. \tag{1.122}$$

The last formula expresses the loss occurring during the monotonic increase of input between u_1 and u_2, $(u_1 > u_2)$:

$$Q(u_1, u_2) = \tfrac{1}{2}[\bar{Q}(u_+, u_2) - \bar{Q}(u_+, u_1)]. \tag{1.123}$$

where u_+ is the last input maximum (see Fig. 1.48).

The formulas (1.122) and (1.123) may be useful from the practical point of view, because it is much easier to measure cyclic losses than those occurring for nonperiodic input variations.

It is instructive to show that the derived expressions for hysteretic energy losses are consistent with the classical result: a hysteretic energy loss occurring for a cyclic input variation is equal to the area enclosed by the loop resulting from this cyclic input variation.

Consider a cyclic input variation between u_- and u_+. According to (1.118) and (1.115), we find that the hysteretic loss per cycle for the above input variation is given by

$$\begin{aligned}\bar{Q}(u_-, u_+) &= 2Q(u_-, u_+)\\ &= 2[(u_+ - u_-)F(u_+, u_-)\\ &\quad - \int_{u_-}^{u_+} F(u_+, \beta)d\beta - \int_{u_-}^{u_+} F(\alpha, u_-)d\alpha].\end{aligned} \tag{1.124}$$

On the other hand, the area enclosed by the corresponding loop is given by

$$W = \oint_{u_-u_+u_-} udf. \tag{1.125}$$

Since

$$\oint_{u_-u_+u_-} udf + \oint_{u_-u_+u_-} fdu = \oint_{u_-u_+u_-} d(uf) = 0, \tag{1.126}$$

we find

$$W = -\oint_{u_-u_+u_-} fdu = -\int_{u_-u_+} fdu - \int_{u_+u_-} fdu. \tag{1.127}$$

To evaluate the last two integrals, we shall use the formulas (1.71) and (1.74). In the case of monotonic input increase, the formula (1.74) is appropriate. This formula can be written in the form

$$f(t) = C + 2F(u(t), u_-), \tag{1.128}$$

where u_- is used instead of m_{n-1} and the constant C is given by

$$C = -F(\alpha_0, \beta_0) + 2\sum_{k=1}^{n-1} [F(M_k, m_{k-1}) - F(M_k, m_k)]. \tag{1.129}$$

From (1.128), we find

$$\int_{u_-u_+} f du = C(u_+ - u_-) + 2\int_{u_-}^{u_+} F(u, u_-)du. \tag{1.130}$$

In the case of monotonic input decrease from u_+ to u_-, the formula (1.71) is appropriate. This formula can be rearranged as

$$f(t) = C + 2[F(u_+, u_-) - F(u_+, u(t))], \tag{1.131}$$

where the notation u_+ is used instead of M_n, and C is the same history-dependent constant as in (1.128).

From (1.131), we obtain

$$\int_{u_+u_-} f du = C(u_- - u_+) + 2(u_- - u_+)F(u_+, u_-) - 2\int_{u_+}^{u_-} F(u_+, u)du. \tag{1.132}$$

Using (1.127), (1.130), and (1.132), we derive

$$W = 2[(u_+ - u_-)F(u_+, u_-) - \int_{u_-}^{u_+} F(u_+, u)du - \int_{u_-}^{u_+} F(u, u_-)du]. \tag{1.133}$$

It is apparent that the expressions (1.124) and (1.133) are identical. This proves that the expressions for hysteretic energy losses derived above are consistent with the classical result (1.125).

We next discuss the applications of the above results to the irreversible thermodynamics of hysteretic media. It was pointed out before that any hysteresis phenomenon is accompanied by energy dissipation. This means that hysteretic processes are irreversible, and consequently they fall in the domain of irreversible thermodynamics. Irreversible thermodynamics is the far-reaching extension of classical thermodynamics which describes reversible processes. For this reason, it is appropriate to begin with a brief review of the formal structure of classical thermodynamics.

Classical thermodynamics is based upon three main principles. The first principle of classical thermodynamics is the law of energy conservation.

According to this principle, there exists a function of state, called the internal energy U of the closed system. This state function is such that its infinitesimally small change dU may occur as a result of energy exchange with the surroundings in the form of heat $đQ$, as well as a result of energy $đW$ added to the system (or spent by the system) in the form of work. Mathematically, the first principle is expressed by

$$dU = đQ + đW \tag{1.134}$$

where a stroke is put across the symbol d in (1.134) to emphasize that the quantities $đQ$ and $đW$ are path-dependent infinitesimals, which are sometimes called imperfect differentials. In order words, the above quantities depend on a particular path traced by the system during its transition from one equilibrium state to another.

The second principle of classical thermodynamics postulates the existence of another state function, called the entropy S. This principle also relates $đQ$ to the differential of entropy dS for reversible processes by the formula

$$dS = \frac{đQ}{T}, \tag{1.135}$$

where T is the absolute temperature.

For irreversible processes, the above equality is replaced by the inequality

$$dS > \frac{đQ}{T}. \tag{1.136}$$

The formulas (1.134) and (1.135) are often combined into one formula:

$$dS = \frac{1}{T}(dU - đW), \tag{1.137}$$

which constitutes the mathematical foundation of classical thermodynamics. The factor $1/T$ in (1.137) and (1.135) can be mathematically interpreted as an integrating factor for the imperfect differential $đQ$.

The first and second principles of classical thermodynamics are complemented by the Nernst-Planck postulate which is sometimes called the third principle of thermodynamics. According to this principle, the entropy of any system vanishes at zero temperature. This principle provides some useful information concerning the asymptotic behavior of entropy. However, the bulk of phenomenological thermodynamics does not require this principle. For this reason, the above postulate can not be compared in importance with the first and second principles of classical thermodynamics.

Classical thermodynamics has by and large been extended in two main directions. The first extension is based on the introduction of new variables describing the composition of the system. This approach has been very successful in applications of thermodynamics to chemical reactions. It has led to the development of chemical thermodynamics. The second extension is based on the generalization of the second principle of classical thermodynamics; it has led to the development of thermodynamic theory of irreversible processes which is often called irreversible thermodynamics. This theory has been mainly developed by I. Prigogine and his collaborators.

As mentioned above, the basic difference between classical and irreversible thermodynamics lies in the way in which the second principle is stated. In irreversible thermodynamics, the second principle is formulated as follows.

1. The change in entropy dS can be split into two parts:

$$dS = d_eS + d_iS, \tag{1.138}$$

where d_eS is due to the flow of entropy into the system from its surroundings, while d_iS is the generation of entropy by irreversible processes within the system. The term d_iS is often called internal entropy production.
2. The entropy production d_iS is never negative. It is zero if only the system undergoes a reversible process and positive if it undergoes an irreversible process. Thus,

$$d_iS \geq 0. \tag{1.139}$$

3. For closed systems, the term d_eS is related to the energy d_eQ received in the form of heat by the formula, which is similar to (1.135),

$$d_eS = \frac{d_eQ}{T}, \tag{1.140}$$

where T is the absolute temperature which is assumed to be definable (at least locally) for nonequilibrium situations.

It is also assumed in irreversible thermodynamics that the formula (1.137) holds with dS meaning the total change in entropy.

From (1.138) and (1.140), we find

$$TdS = d_eQ + Td_iS. \tag{1.141}$$

The last expression allows one to relate the entropy production to the dissipated energy within the system. Indeed, from (1.137) and (1.141) we find

$$dU = dW + d_eQ + Td_iS. \tag{1.142}$$

The first two terms in (1.142) have the same meaning as in (1.134) while the last term can be interpreted as the energy supplied to the system as a result of dissipating (irreversible) processes. This interpretation is especially clear for adiabatically isolated ($d_eQ = 0$) systems which undergo cyclic changes ($\oint dW = 0$). The increase in internal energy during one cycle ($\oint dU$) of such systems is only due to dissipating processes within the system ($\oint Td_iS$). From (1.142) and the foregoing discussion, we find that the entropy production can be related to the dissipated energy d_iQ by the formula

$$d_iS = \frac{d_iQ}{T}. \tag{1.143}$$

It is apparent that the novel part of the above formulation of the second principle is the introduction of internal entropy production d_iS. However, this new quantity is useful if only it can be evaluated (in a mathematical form) for different irreversible processes. This is the central problem of irreversible thermodynamics and it has been emphasized in many books on this subject. For instance, in the book [40] by I. Prigogine we find (pp. 90): "The main feature of the thermodynamics of irreversible processes consists of the evaluation of the entropy productions ...." Similarly, in [41] we read (pp. 21): "In thermodynamics of irreversible procees, however, one of the important objectives is to relate the quantity d_iS, the entropy production, to the various irreversible phenomena which may occur inside the system." Finally, in [42] it is noted (pp. 69): "A central problem of irreversible thermodynamics, in fact, is the development of formulas for the entropy production d_iS in specific cases."

The above problem has been resolved for various irreversible processes which are caused by *macroscopic nonuniformities* of the system. Examples of such processes include heat flows due to temperature gradients, diffusion flows due to density gradients, electric current flows due to potential gradients, and so forth. For the above processes, the so-called entropy balance equation (1.138) is written in local (differential) form in which the entropy production d_iS is replaced by the entropy source. The entropy source is then found as a sum of several terms each being the product of a flux characterizing a particular irreversible process and a quantity, called thermodynamic force, which is related to a particular macrospic nonuniformity of the system (temperature gradient, for instance). In this way many useful results have been established. The celebrated Onsanger reciprocity principle for phenomenological coefficients is the most known example of these results. However, the above developments can not cover hysteresis phenomena. This is because hysteresis is not caused by macroscopic nonuniformities and therefore can not be linked to gradients of some physical quantities. As a result, different approches should be developed for the calculation of entropy production in the case of irreversible hysteretic processes. It is logical to expect that mathematical models of hysteresis may help to solve the above problem. It is shown below that by using the Preisach model and the expressions for hysteretic energy losses derived on the basis of this model, the entropy production for hysteresis processes can indeed be found.

Consider the input $u(t)$ which is monotonically increased from its previous local minimum value u_-. If the current input value $u(t)$ is lower than M_n, then by using (1.115) for the dissipated energy we find

$$Q(u_-, u) = (u - u_-)F(u, u_-) - \int_{u_-}^{u} F(u, \beta)d\beta - \int_{u_-}^{u} F(\alpha, u_-)d\alpha. \quad (1.144)$$

Employing (1.144), it is easy to conclude that the energy dissipation d_iQ, which occurs as a result of the input increase from u to $u + du$, is given by

$$d_i Q = \frac{\partial Q(u_-, u)}{\partial u} du. \tag{1.145}$$

From (1.145) and (1.143), we derive

$$d_i S = \frac{1}{T} \frac{\partial Q(u_-, u)}{\partial u} du. \tag{1.146}$$

By differentiating (1.144) with respect to u and substituting the result into (1.146), after simple transformations we obtain

$$d_i S = \frac{du}{T} \int_{u_-}^{u} \frac{\partial}{\partial u} [F(u, u_-) - F(u, \beta)] d\beta. \tag{1.147}$$

If the input is monotonically decreased from its previous local maximum value, then the entropy production occurring as a result of the input decrease from u to $u - du$ is given by

$$d_i S = \frac{du}{T} \int_{u}^{u_+} \frac{\partial}{\partial u} [F(u_+, u) - F(\alpha, u)] d\alpha. \tag{1.148}$$

The derivation of the last formula is similar to that for (1.147). It has been tacitly assumed in the previous derivation that the monotonic input variations are such that no previous history is wiped out. However, by using (1.116), it is easy to extend (1.147) and (1.148) to the most general case. The details of this extension are left to the reader.

6 Preisach Model with Stochastic Input as a Model for Viscosity

It is well known that the physical origin of hystereis is due to the multiplicity of metastable states exhibited by hysteretic materials or systems. At equilibrium, large deviations of random (thermal) perturbations may cause a hysteretic system to move from higher to lower energy metastable states. This may result in gradual (slow) changes of the output variable. This temporal loss of memory of a hysteretic system is generally referred to in the literature as "after effect," "viscosity" or "creep." This phenomenon can be of practical significance in various engineering applications where hysteresis is utilized. One important example is the magnetic storage technology where the viscosity effect is important as far as the time reliability of recorded information is concerned.

Traditionally, the viscosity phenomenon has been studied by using thermal activation type models (see, for example, [44, 68]) It has been realized that these models have some intrinsic limitations. For instance, they are usually regarded as "noninteracting particle" models and they are valid for hysteretic systems with energy barriers much larger than the thermal activation energy. In this section we shall explore a new approach to the modelling of the viscosity phenomenon based upon the Preisach model of hysteresis. In this approach random thermal agitations will be modelled by a stochastic input. We shall compare this approach with thermal activation-type models and with some known experimental facts.

Consider a deterministic input $u(t)$ which at time $t = 0$ assumes some value u and remains constant thereafter. In a purely deterministic situation, the output $f(t)$ would remain constant for $t \geq 0$ as well. However, in order to model the viscosity phenomenon, we assume that some noise described by stochastic process X_t is superimposed on the constant input u. Consequently, the Preisach model is driven by the following random process:

$$x_t = u + X_t, \qquad \bar{X}_t = 0, \tag{1.149}$$

where $\bar{X}_t$ stands for the expected value of X_t.

The output will also be a random process given by

$$f_t = \iint_{\alpha \geq \beta} \mu(\alpha, \beta)\hat{\gamma}_{\alpha\beta} x_t \, d\alpha \, d\beta. \tag{1.150}$$

The hysteresis operators $\hat{\gamma}_{\alpha\beta}$ are naturally defined for piecewise monotonic inputs. However, quite often samples of stochastic processes can not be described in terms of piecewise monotonic functions. This creates some mathematical difficulties. To get around these and some other difficulties, and the mathematical subtleties associated with them, we shall model the noise by a discrete-time independent identically distributed (i.i.d.) random process X_n. In this case, it can be assumed that process samples remain constant at time intervals $\Delta t = t_{n+1} - t_n$ and undergo monotonic step changes at times t_{n+1}. Accordingly, the Preisach model is driven by the process

$$x_n = u + X_n, \qquad \bar{X}_n = 0. \tag{1.151}$$

The output process is given by

$$f_n = \iint_{\alpha \geq \beta} \mu(\alpha, \beta)\hat{\gamma}_{\alpha\beta} x_n \, d\alpha \, d\beta. \tag{1.152}$$

We will be interested in the time evolution of the expected value of the output process. Since integration is a linear operation, from (1.152) we derive

$$\bar{f}_n = E\{f_n\} = \iint_{\alpha \geq \beta} \mu(\alpha, \beta) E\{\hat{\gamma}_{\alpha\beta} x_n\} \, d\alpha \, d\beta. \tag{1.153}$$

Thus, the whole problem is reduced to the evaluation of the expected value, $E\{\hat{\gamma}_{\alpha\beta} x_n\}$. Since $\hat{\gamma}_{\alpha\beta} x_n$ may assume only two values $+1$ and -1, we find

$$E\{\hat{\gamma}_{\alpha\beta} x_n\} = P\{\hat{\gamma}_{\alpha\beta} x_n = +1\} - P\{\hat{\gamma}_{\alpha\beta} x_n = -1\}. \tag{1.154}$$

It is also clear that

$$P\{\hat{\gamma}_{\alpha\beta} x_n = +1\} + P\{\hat{\gamma}_{\alpha\beta} x_n = -1\} = 1. \tag{1.155}$$

By introducing the notation

$$P\{\hat{\gamma}_{\alpha\beta}x_n = +1\} = q_{\alpha\beta}(n), \tag{1.156}$$

from (1.154) and (1.155) we obtain

$$E\{\hat{\gamma}_{\alpha\beta}x_n\} = 2q_{\alpha\beta}(n) - 1. \tag{1.157}$$

We next derive the finite difference equation for $q_{\alpha\beta}(n)$. According to the total probability theorem, we have

$$\begin{aligned} P\{\hat{\gamma}_{\alpha\beta}x_{n+1} = +1\} &= P\{\hat{\gamma}_{\alpha\beta}x_{n+1} = +1|\hat{\gamma}_{\alpha\beta}x_n = +1\}P\{\hat{\gamma}_{\alpha\beta}x_n = +1\} \\ &\quad + P\{\hat{\gamma}_{\alpha\beta}x_{n+1} = +1|\hat{\gamma}_{\alpha\beta}x_n = -1\}P\{\hat{\gamma}_{\alpha\beta}x_n = -1\}. \end{aligned} \tag{1.158}$$

It is convenient now to introduce the switching probabilities

$$P_{\alpha\beta}^{++}(n) = P\{\hat{\gamma}_{\alpha\beta}x_{n+1} = +1|\hat{\gamma}_{\alpha\beta}x_n = +1\}, \tag{1.159}$$

$$P_{\alpha\beta}^{-+}(n) = P\{\hat{\gamma}_{\alpha\beta}x_{n+1} = +1|\hat{\gamma}_{\alpha\beta}x_n = -1\}. \tag{1.160}$$

Similar meanings hold for the switching probabilities $P_{\alpha\beta}^{--}(n)$ and $P_{\alpha\beta}^{+-}(n)$. It is clear that

$$P_{\alpha\beta}^{++}(n) + P_{\alpha\beta}^{+-}(n) = 1, \tag{1.160}$$

$$P_{\alpha\beta}^{-+}(n) + P_{\alpha\beta}^{--}(n) = 1, \tag{1.161}$$

$$P\{\hat{\gamma}_{\alpha\beta}x_n = -1\} = 1 - q_{\alpha\beta}(n). \tag{1.162}$$

By using (1.156) and (1.159)–(1.162), we can transform (1.158) to

$$q_{\alpha\beta}(n+1) = q_{\alpha\beta}(n)[1 - (P_{\alpha\beta}^{-+}(n) + P_{\alpha\beta}^{+-}(n))] + P_{\alpha\beta}^{-+}(n). \tag{1.163}$$

We next proceed to the evaluation of switching probabilities $P_{\alpha\beta}^{-+}(n)$ and $P_{\alpha\beta}^{+-}(n)$. It is clear that these probabilities can also be defined as

$$P_{\alpha\beta}^{-+}(n) = P\{x_{n+1} > \alpha|\hat{\gamma}_{\alpha\beta}x_n = -1\}. \tag{1.164}$$

$$P_{\alpha\beta}^{+-}(n) = P\{x_{n+1} < \beta|\hat{\gamma}_{\alpha\beta}x_n = +1\}. \tag{1.165}$$

In general, these probabilities are difficult to evaluate because multidimensional conditional probability density functions are required. However, the problem is significantly simplified if the noise is modelled by i.i.d. process. In this case we find

$$P_{\alpha\beta}^{-+}(n) = P_{\alpha}^{-+} = P\{x_{n+1} > \alpha\} = \int_{\alpha}^{\infty} \rho(x)dx, \tag{1.166}$$

$$P_{\alpha\beta}^{+-}(n) = P_{\beta}^{+-} = P\{x_{n+1} > \beta\} = \int_{-\infty}^{\beta} \rho(x)dx, \tag{1.167}$$

where $\rho(x)$ is a probability density function.

By using (1.166) and (1.167), the finite difference equation (1.163) can be represented as

$$q_{\alpha\beta}(n+1) = r_{\alpha\beta} q_{\alpha\beta}(n) + P_\alpha^{-+}, \tag{1.168}$$

where

$$r_{\alpha\beta} = 1 - (P_\alpha^{-+} + P_\beta^{+-}) = \int_\beta^\alpha \rho(x)dx. \tag{1.169}$$

If the probability density function $\rho(x)$ is strictly positive, then

$$0 < r_{\alpha\beta} < 1. \tag{1.170}$$

Equation (1.168) is a constant coefficient first-order finite difference equation whose general solution has the form

$$q_{\alpha\beta}(n) = A r_{\alpha\beta}^n + B. \tag{1.171}$$

By substituting (1.171) into (1.168), we find

$$B = \frac{P_\alpha^{-+}}{1 - r_{\alpha\beta}} = \frac{P_\alpha^{-+}}{P_\alpha^{-+} + P_\beta^{+-}}. \tag{1.172}$$

From the initial condition we obtain

$$q_{\alpha\beta}(0) = A + B, \tag{1.173}$$

where

$$q_{\alpha\beta}(0) = \begin{cases} 1, & \text{if } (\alpha, \beta) \in S^+(0), \\ 0, & \text{if } (\alpha, \beta) \in S^-(0), \end{cases} \tag{1.174}$$

and $S^+(0)$ and $S^-(0)$ are positive and negative sets on the $\alpha - \beta$ diagram, respectively, at the instant of time $t = 0$, From (1.171)–(1.173), we derive

$$q_{\alpha\beta}(n) = [q_{\alpha\beta}(0) - q_{\alpha\beta}(\infty)] r_{\alpha\beta}^n + \frac{P_\alpha^{-+}}{P_\alpha^{-+} + P_\beta^{+-}}, \tag{1.175}$$

where

$$q_{\alpha\beta}(\infty) = \lim_{n\to\infty} q_{\alpha\beta}(n) = \frac{P_\alpha^{-+}}{P_\alpha^{-+} + P_\beta^{+-}}. \tag{1.176}$$

By substituting (1.175) into (1.157), we obtain

$$E\{\hat{\gamma}_{\alpha\beta} x_n\} = 2[q_{\alpha\beta}(0) - q_{\alpha\beta}(\infty)] r_{\alpha\beta}^n + \frac{P_\alpha^{-+} - P_\beta^{+-}}{P_\alpha^{-+} + P_\beta^{+-}}. \tag{1.177}$$

We next introduce the functions

$$\zeta(\alpha, \beta) = \frac{P_\alpha^{-+} - P_\beta^{+-}}{P_\alpha^{-+} + P_\beta^{+-}}, \tag{1.178}$$

$$\vartheta(\alpha, \beta) = \begin{cases} 1, & \text{if } (\alpha, \beta) \in S^+(0), \\ -1, & \text{if } (\alpha, \beta) \in S^-(0). \end{cases} \tag{1.179}$$

By using (1.174), (1.178), and (1.179), we can transform (1.177) to

$$E\{\hat{\gamma}_{\alpha\beta}x_n\} = [\vartheta(\alpha, \beta) - \zeta(\alpha, \beta)]r_{\alpha\beta}^n + \zeta(\alpha, \beta). \tag{1.180}$$

By substituting (1.180) into (1.153), we finally find

$$\bar{f}_n = \bar{f}_\infty + \iint_{\alpha \geq \beta} \mu(\alpha, \beta)[\vartheta(\alpha, \beta) - \zeta(\alpha, \beta)]r_{\alpha\beta}^n d\alpha d\beta, \tag{1.181}$$

where

$$\bar{f}_\infty = \lim_{n \to \infty} \bar{f}_n = \iint_{\alpha \geq \beta} \mu(\alpha, \beta)\zeta(\alpha, \beta) d\alpha d\beta. \tag{1.182}$$

It is apparent from (1.182) and (1.178) that the limiting expected value of output $\bar{f}_\infty$ does not depend on the history of input variations prior to the time $t = 0$. In this respect, the value $\bar{f}_\infty$ bears some resemblance to anhysteretic output value. This resemblance is enhanced by the fact that $\bar{f}_\infty = 0$ if the expected value u of x_n is equal to zero, and $\rho(x) = \rho(X)$ is an even function. This fact can be proven as follows. According to (1.166) and (1.167), we find

$$P_\alpha^{-+} = P_{-\alpha}^{+-}. \tag{1.183}$$

By using (1.183), it is easy to check that

$$\zeta(\alpha, \beta) = -\zeta(-\beta, -\alpha). \tag{1.184}$$

On the other hand, we recall that

$$\mu(\alpha, \beta) = \mu(-\beta, -\alpha). \tag{1.185}$$

From (1.184) and (1.185), we conclude that $\mu(\alpha, \beta)\zeta(\alpha, \beta)$ is an odd function with respect to the line $\alpha = -\beta$. From the last fact and (1.182), we find that

$$\bar{f}_\infty = 0. \tag{1.186}$$

Next, we shall compare the result (1.181) with thermal activation-type models for viscosity. For this purpose, we replace discrete time n by continuous time t and rewrite (1.181) as

$$\bar{f}_t = \bar{f}_\infty + \iint_{\alpha \geq \beta} \chi(\alpha, \beta)e^{-\xi(\alpha, \beta)t} d\alpha d\beta, \tag{1.187}$$

where

$$\chi(\alpha, \beta) = \mu(\alpha, \beta)[\vartheta(\alpha, \beta) - \zeta(\alpha, \beta)], \tag{1.188}$$

$$\xi(\alpha, \beta) = -\frac{1}{\Delta t} \ln r_{\alpha\beta}. \tag{1.189}$$

According to thermal activation-type models, metastable equilbrium states of hysteretic systems are separated by energy barriers E_B. It is assumed that there is a continuum of these energy barriers and it is postulated (with some physical justification) that the viscosity phenomenon is described by the model (see [44]):

$$f(t) = f(\infty) + A \int_0^\infty g(E_B) e^{-\lambda(E_B)t} \, dE_B, \tag{1.190}$$

where $g(E_B)$ is some density of states,

$$\lambda(E_B) = \lambda_0 e^{E_B/kT}, \tag{1.191}$$

k is Boltzman's constant, T is the absolute temperature, and A and λ_0 are some constants.

It is clear by inspection that there is some similarity between our result (1.187) and the thermal activation model (1.190). Acutally, our model (1.187) can be reduced to (1.190) in the particular case when only symmetrical loops (operators) $\hat{\gamma}_{\alpha,-\alpha}$ are used in the Preisach model. In this case

$$\mu(\alpha, \beta) = \theta(\alpha, \beta)\delta(\alpha + \beta), \tag{1.192}$$

where $\delta(\alpha + \beta)$ is the Dirac-delta function.

By substituting (1.192) into (1.188) and (1.187), after simple transformations we can represent (1.187) as

$$\bar{f}_t = \bar{f}_\infty + \int_0^\infty \tilde{\chi}(\alpha) e^{-\tilde{\xi}(\alpha)t} \, d\alpha. \tag{1.193}$$

Now, by using the change of variables

$$\tilde{\xi}(\alpha) = \lambda(E_B), \qquad \alpha = \tilde{\xi}^{-1}(\lambda(E_B)), \tag{1.194}$$

the expression (1.193) can be reduced to (1.190). This shows that the Preisach model of viscosity (1.187) is reduced to the thermal activation model in a very particular case. This case occurs when only symmetrical rectangular loops are used in the Preisach model. Since it is generally believed that nonsymmetrical loops in the Preisach model account for "particle interactions," the last reduction is consistent with the generally held opinion that the thermal activation model (1.190) is a "noninteracting particle" model.

We next show that under some "traditional" assumptions about $\chi(\alpha, \beta)$ the model (1.187) describes ln t-type variations for $\bar{f}_t$. The ln t-type variations have been observed in many experiments and they are considered to be characteristic for viscosity phenomena.

The expression (1.187) can be written as

$$\bar{f}_t = \bar{f}_\infty + \int \left(\int \chi(\alpha, \beta) e^{-\xi(\alpha,\beta)t} \, d\alpha \right) d\beta. \tag{1.195}$$

For any fixed β, we make the change of variables from α to ξ in the integral with respect to α:

$$\bar{f}_t = \bar{f}_\infty + \int \left(\int \psi(\xi, \beta) e^{-\xi t} d\xi \right) d\beta. \tag{1.196}$$

We assume that the function $\psi(\xi, \beta)$ can be approximated as

$$\psi(\xi, \beta) \approx \frac{D(\beta)}{\xi} \quad \text{for} \quad \xi_1 < \xi < \xi_2 \tag{1.197}$$

and zero otherwise, Then, we derive

$$\int \psi(\xi, \beta) e^{-\xi t} d\xi \approx D(\beta) \int_{\xi_1}^{\xi_2} \frac{e^{-\xi t}}{\xi} d\xi = D(\beta) \int_{\xi_1 t}^{\xi_2 t} \frac{e^{-v}}{v} dv. \tag{1.198}$$

Consider such instants of time t that

$$\frac{1}{\xi_2} \ll t \ll \frac{1}{\xi_1}. \tag{1.199}$$

For these instants of time, from (1.198) we derive

$$\begin{aligned} \int \psi(\xi, \beta) e^{-\xi t} d\xi &\approx D(\beta) \int_{\xi_1 t}^{\infty} \frac{e^{-v}}{v} dv \\ &= D(\beta) \left[\int_{-a}^{\infty} \frac{e^{-v}}{v} dv - \int_{-a}^{\xi_1 t} \frac{e^{-v}}{v} dv \right]. \end{aligned} \tag{1.200}$$

Here a is a fixed small positive number and the last two integrals exist in the sense of Cauchy principal value. Since $\xi_1 t \ll 1$, we can use only the first two terms of Taylor expansion of e^{-v} in the second integral. This yields

$$\int \psi(\xi, \beta) e^{-\xi t} d\xi \approx S(\beta) \ln t + C(\beta). \tag{1.201}$$

By substituting (1.201) into (1.196), we find

$$\bar{f}_t \approx S \ln t + C, \tag{1.202}$$

which is a famous ln t-type variation for $\bar{f}_t$.

It has been observed in experiments (see, for instance, [69]) that the temperature dependence of the decay rate of $\bar{f}_t$ is satisfactorily approximated by $T^{1/2}$ at low temperatures. In the framework of the model (1.187) this fact can be explained as follows. From (1.189) and (1.169), we find

$$\xi(\alpha, \beta) = \frac{-1}{\Delta t} \ln \left[\int_\beta^\alpha \rho(x) dx \right]. \tag{1.203}$$

We subdivide all exponents in (1.187) into two groups: (a) exponents for such α's and β's that $u = \bar{x}_t \notin [\beta, \alpha]$, and (b) exponents for such α's and β's that $u = \bar{x}_t \in [\beta, \alpha]$. Consider a thermal noise with small variance σ^2. If $u = \bar{x}_t \notin [\beta, \alpha]$, then

$$\int_\beta^\alpha \rho(x) dx \to 0 \quad \text{as} \quad \sigma \to 0. \tag{1.204}$$

This means according to (1.203) that the exponents of the first group decay very fast and they do not contribute appreciably to the "long time" decay of the integral in (1.187).

If $u = \bar{x}_t \in [\beta, \alpha]$, then

$$\int_{\beta}^{\alpha} \rho(x)dx \to 1 \quad \text{as} \quad \sigma \to 0. \tag{1.205}$$

This means that the exponents of the second group decay slowly and they are by and large responsible for the "long time" decay of $\bar{f}_t$. For this reason, only these exponents are discussed below. It is apparent that

$$\int_{\beta}^{\alpha} \rho(x)dx = 1 - \int_{\beta}^{\infty} \rho(x)dx - \int_{-\infty}^{\beta} \rho(x)dx. \tag{1.206}$$

For small σ the integrals in (1.206) are small quantities which we assume can be approximated (not necessarily asymptotically) as

$$\int_{\alpha}^{\infty} \rho(x)dx \approx a(\alpha)\sigma, \qquad \int_{-\infty}^{\beta} \rho(x)d \approx b(\beta)\sigma. \tag{1.207}$$

By substituting (1.207) into (1.206) and then into (1.203), we find

$$\xi(\alpha, \beta) \approx \frac{-1}{\Delta t} \ln\{1 - [a(\alpha) + b(\beta)]\sigma\} \approx \frac{1}{\Delta t}[a(\alpha) + b(\beta)]\sigma. \tag{1.208}$$

It is reasonable to assume on the physical grounds that the variance is proportional to the temperature:

$$\sigma^2 \sim T. \tag{1.209}$$

From (1.208) and (1.209) we conclude that, for the exponents of the second group, the rate of decay is approximated as

$$\xi(\alpha, \beta) \sim T^{1/2}. \tag{1.210}$$

This suggests that the rate of "long time" decay of $\bar{f}_t$ may be approximated by $T^{1/2}$ at low temperatures.

It is also interesting to examine the dependence of the limiting expected value of the output $\bar{f}_\infty$ on the variance σ^2 of the noise. Expression (1.182) is to be used for this purpose. The noise can be chosen to have normal distribution and $\xi(\alpha, \beta)$ in (1.182) can be computed by using formulas (1.178), (1.166), and (1.167). For illustrative purposes, Preisach function $\mu(\alpha, \beta)$ can be chosen as

$$\mu(\alpha, \beta) = A \exp\left\{-\left[\frac{(\alpha+\beta)^2}{4\sigma_1^2} + \frac{(\alpha - \beta - 2u_c)^2}{4\sigma_2^2}\right]\right\}, \tag{1.211}$$

where A is some normalization constant, u_c is the coercive value of a major hysteretic loop, while σ_1 and σ_2 are parameters which are used to control the spread of the function $\mu(\alpha, \beta)$ along and around the line $\alpha = -\beta$. This Gausian

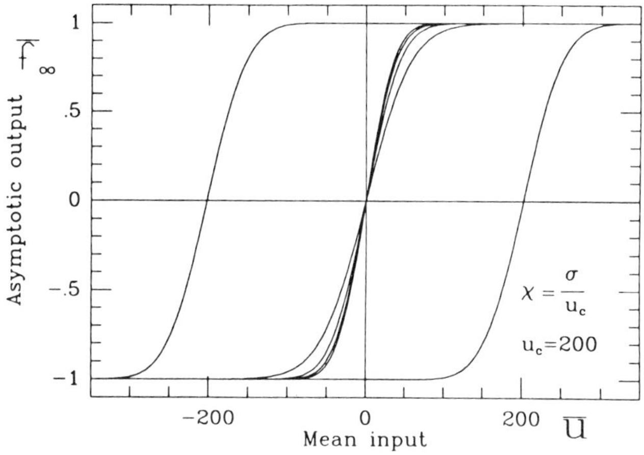

Fig. 1.49

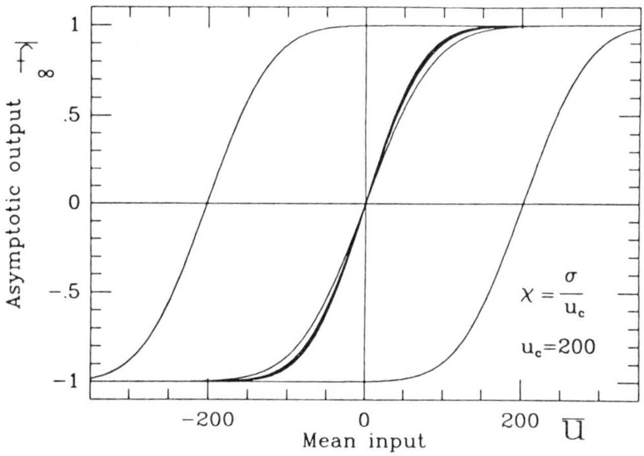

Fig. 1.50

form of $\mu(\alpha, \beta)$ is often used in magnetics [21]. The computations have been performed for different values of σ_1, σ_2, and the ratio $\chi = \sigma/u_c$. The computations show that for arbitrary fixed values of σ_1 and σ_2 the dependence of $\bar{f}_\infty$ on u is not sensitive to wide variations of σ. This is demonstrated in Fig. 1.49 and 1.50 where the computed curves $\bar{f}_\infty$ vs u are plotted for the following values of χ: 0.5, 0.35, 0.2, 0.15. These curves were computed for the following values of σ_1 and σ_2, respectively: $\sigma_1 = \sigma_2 = 40$, $\sigma_1 = 80$, and $\sigma_2 = 40$. The fact that $\bar{f}_\infty$ is insensitive to the variance σ^2 suggests that $\bar{f}_\infty$ can probably be used as another definition of anhysteretic value.

CHAPTER 2

Generalized Scalar Preisach Models of Hysteresis

1 Moving Preisach Model of Hysteresis

The classical Preisach model of hysteresis has been discussed in detail in the previous chapter. It has repeatedly been emphasized that this model has some intrinsic limitations. The most important of them are the following:

1. The classical Preisach model describes hysteresis nonlinearities which exhibit congruency of minor loops formed for the same reversal values of input. However, many experiments show that actual hysteresis nonlinearities may substantially deviate from this property.
2. The classical Preisach model is static in nature and does not account for dynamic properties of hysteresis nonlinearities. However, for fast input variations these properties may be essential.
3. The classical Preisach model describes hysteresis nonlinearities with the wiping-out property. This property is tantamount to the immediate formation of hysteresis loop after only one cycle of back-and-forth variation of input between any two reversal values. However, experiments show that hysteresis loop formation is often preceded by some stabilization process which may require large number of cycles to achieve a stable minor loop. This process is also called in the literature "accommodation" or "reptation" process.
4. The classical Preisach model deals only with scalar hysteresis nonlinearities. However, in many applications vector hysteresis is encountered. Properties of this hysteresis are usually quite different from scalar hysteresis properties.

To remove (or relax) the above-mentioned limitations, essential generalizations of the classical Preisach model are needed. These generalizations for the case of scalar hysteresis are discussed in this chapter, while vector Preisach models are treated in the Chapter III. In this chapter we shall also discuss experimental testing of different Preisach-type models of hysteresis as well as their application to the description of superconducting hysteresis. The latter topic may be of special importance due to the strong interest in superconductivity rekindled by recent remarkable discoveries of high-temperature superconductors.

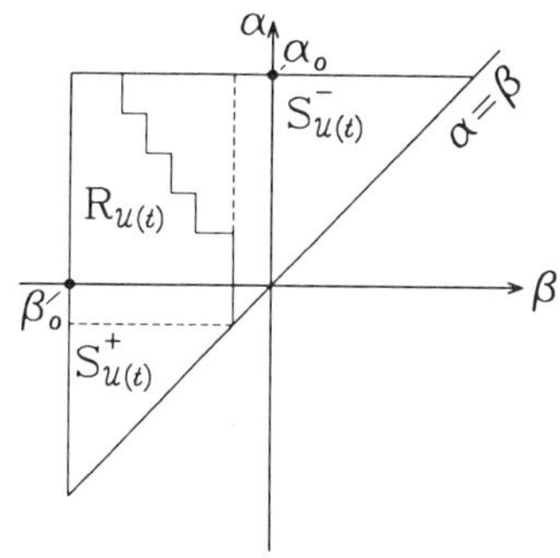

Fig. 2.1

We begin this chapter with some interesting modification of the classical Preisach model. This modification will reveal that the Preisach model does describe to a certain extent reversible properties of hysteresis nonlinearities. This fact has been overlooked in the existing literature. Apart from the mentioned fact, this modification will be also instrumental in the further generalizations of the classical Preisach model which are discussed in subsequent sections.

The classical Preisach model has been defined as

$$f(t) = \hat{\Gamma}u(t) = \iint_T \mu(\alpha, \beta)\hat{\gamma}_{\alpha\beta}u(t)d\alpha d\beta, \tag{2.1}$$

where T is the limiting triangle specified by inequalities $\beta_0 \leq \beta \leq \alpha \leq \alpha_0$. This triangle is the support of the function $\mu(\alpha, \beta)$, and it does not change with input variations.

We next subdivide the triangle T into three sets $S^+_{u(t)}$, $R_{u(t)}$, and $S^-_{u(t)}$ (see Fig. 2.1), which are defined as

$$(\alpha, \beta) \in S^+_{u(t)}, \quad \text{if} \quad \beta_0 \leq \beta \leq \alpha \leq u(t), \tag{2.2}$$

$$(\alpha, \beta) \in R_u(t), \quad \text{if} \quad \beta_0 \leq \beta \leq u(t), \quad u(t) \leq \alpha \leq \alpha_0, \tag{2.3}$$

$$(\alpha, \beta) \in S^-_{u(t)}, \quad \text{if} \quad u(t) \leq \beta \leq \alpha \leq \alpha_0. \tag{2.4}$$

By using the above subdivision, we can represent (2.1) in the form

$$f(t) = \iint_{R_{u(t)}} \mu(\alpha, \beta)\hat{\gamma}_{\alpha\beta}u(t)d\alpha d\beta + \iint_{S^+_{u(t)}} \mu(\alpha, \beta)\hat{\gamma}_{\alpha\beta}u(t)d\alpha d\beta$$
$$+ \iint_{S^-_{u(t)}} \mu(\alpha, \beta)\hat{\gamma}_{\alpha\beta}u(t)d\alpha d\beta. \tag{2.5}$$

Since $u(t) \geq \alpha$ for any $(\alpha, \beta) \in S^+_{u(t)}$, then $\hat{\gamma}_{\alpha\beta}u(t) = +1$ and

$$\iint_{S^+_{u(t)}} \mu(\alpha, \beta)\hat{\gamma}_{\alpha\beta}u(t)d\alpha d\beta = \iint_{S^+_{u(t)}} \mu(\alpha, \beta)d\alpha d\beta. \tag{2.6}$$

Similarly, $u(t) \leq \beta$ for any $(\alpha, \beta) \in S^-_{u(t)}$ and

$$\iint_{S^-_{u(t)}} \mu(\alpha, \beta)\hat{\gamma}_{\alpha\beta}u(t)d\alpha d\beta = -\iint_{S^-_{u(t)}} \mu(\alpha, \beta)d\alpha d\beta. \tag{2.7}$$

By substituting (2.6) and (2.7) into (2.5), we find

$$f(t) = \iint_{R_{u(t)}} \mu(\alpha, \beta)\hat{\gamma}_{\alpha\beta}u(t)d\alpha d\beta + \iint_{S^+_{u(t)}} \mu(\alpha, \beta)d\alpha d\beta - \iint_{S^-_{u(t)}} \mu(\alpha, \beta)d\alpha d\beta. \tag{2.8}$$

We next find a simple expression for the last two terms in (2.8). Consider a monotonic increase of input from some value below β_0 (state of negative saturation) to some value $u(t)$. Then the output will change along the ascending branch $f^+_{u(t)}$, and according to (2.8) we find

$$f^+_{u(t)} = -\iint_{R_{u(t)}} \mu(\alpha, \beta)d\alpha d\beta + \iint_{S^+_{u(t)}} \mu(\alpha, \beta)d\alpha d\beta - \iint_{S^-_{u(t)}} \mu(\alpha, \beta)d\alpha d\beta. \tag{2.9}$$

Similarly, if we consider a monotonic decrease of input from some value above α_0 (state of positive saturation) to the same value $u(t)$, then by using (2.8) we obtain

$$f^-_{u(t)} = \iint_{R_{u(t)}} \mu(\alpha, \beta)d\alpha d\beta + \iint_{S^+_{u(t)}} \mu(\alpha, \beta)d\alpha d\beta - \iint_{S^-_{u(t)}} \mu(\alpha, \beta)d\alpha d\beta. \tag{2.10}$$

By summing up (2.9) and (2.10), we derive

$$\iint_{S^+_{u(t)}} \mu(\alpha, \beta)d\alpha d\beta - \iint_{S^-_{u(t)}} \mu(\alpha, \beta)d\alpha d\beta = \frac{1}{2}(f^+_{u(t)} + f^-_{u(t)}). \tag{2.11}$$

By substituting (2.11) into (2.8), we finally obtain

$$f(t) = \iint_{R_{u(t)}} \mu(\alpha, \beta)\hat{\gamma}_{\alpha\beta}u(t)d\alpha d\beta + \frac{1}{2}(f^+_{u(t)} + f^-_{u(t)}). \tag{2.12}$$

The last expression is formally equivalent to the classical one (2.1). However, in this expression the integration is performed not over the fixed limiting triangle T but over the rectangle $R_{u(t)}$ which changes along with input variations. For this reason, the expression (2.12) is termed here as a moving Preisach model. It is also clear from (2.12) that term $\frac{1}{2}(f^+_{u(t)} + f^-_{u(t)})$ represents a fully reversible component of hysteresis nonlinearity described by the classical Preisach model. In this respect, the first term in the right-hand side of (2.12) can be construed as an irreversible component of the classical Preisach model. To make the last point transparent, consider some output increment Δf_u corresponding to some input value u (see Fig. 2.2). This increment depends on a particular history of input variations and can be regarded as a measure of

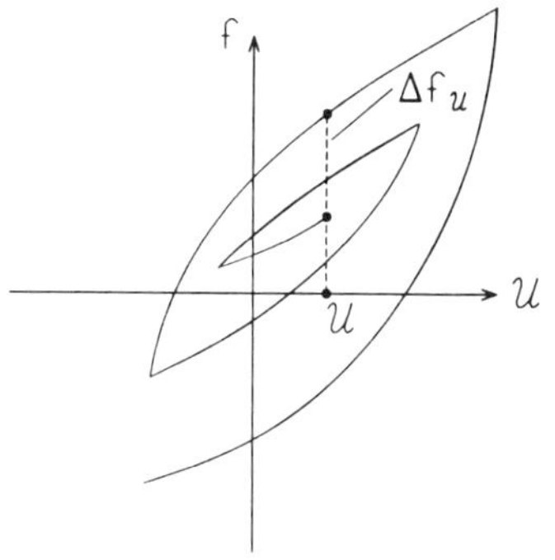

Fig. 2.2

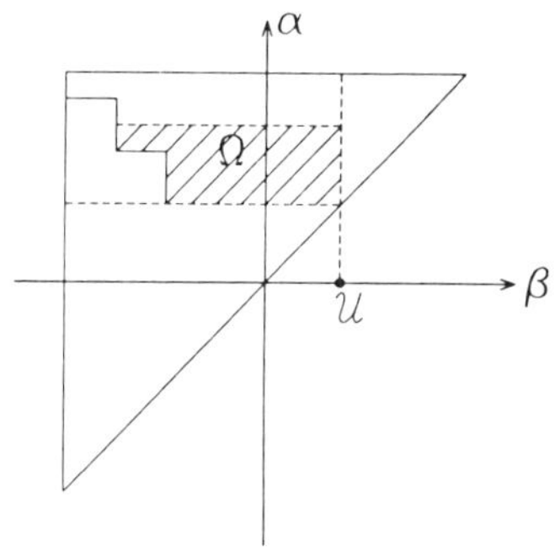

Fig. 2.3

irreversibility of hysteresis nonlinearity for this particular history. It is clear from the diagram shown in Fig. 2.3 and the expression (2.12) that the above increment is given by

$$\Delta f_u = \iint_{\Omega} \mu(\alpha, \beta) d\alpha d\beta. \tag{2.13}$$

The region of integration Ω belongs to $R_{u(t)}$. This is true for any past history. This fact clearly suggests that the first term in the right-hand side of (2.12) describes irreversible processes. Thus, the expression (2.12) gives the decomposition of hysteresis nonlinearity described by the classical Preisach model into irreversible and reversible components.

It is stressed in the above discussion that the expression (2.12) is mathematically equivalent to the classical definition (2.1). However, it is apparent that this equivalence holds only for input and output variations confined to the region enclosed by a major hysteresis loop. Outside this region, the classical Preisach model prescribes flat saturation values for output, while the moving model (2.12) prescribes the actual experimentally observed values $f^+_{u(t)}$ and $f^-_{u(t)}$ for the states of negative and positive saturation, respectively. This is the case because for these states ascending and descending branches merge together and consequently $f^+_{u(t)} = f^-_{u(t)}$. As far as the first term in (2.12) is concerned, it is clear from (2.3) that this term vanishes. Thus, the moving model (2.12) can be regarded as a generalization of the classical model as far as the description of hysteresis nonlinearities beyond the limits of major loops is concerned.

It is instructive to consider directly the identification problem for the model (2.12) without invoking our previous discussion of this problem for the classical model (2.1). The essence of this problem is in determining the function μ by fitting the model (2.12) to some experimental data. Suppose that, starting from the state of negative saturation, the input $u(t)$ is monotonically increased until it reaches some value α. Then the input is monotonically decreased until

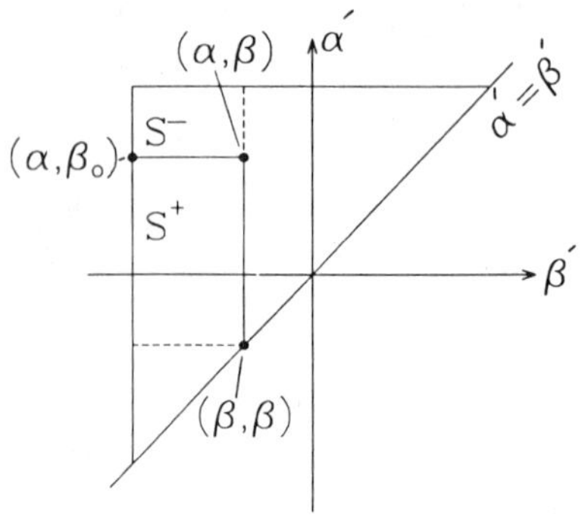

Fig. 2.4

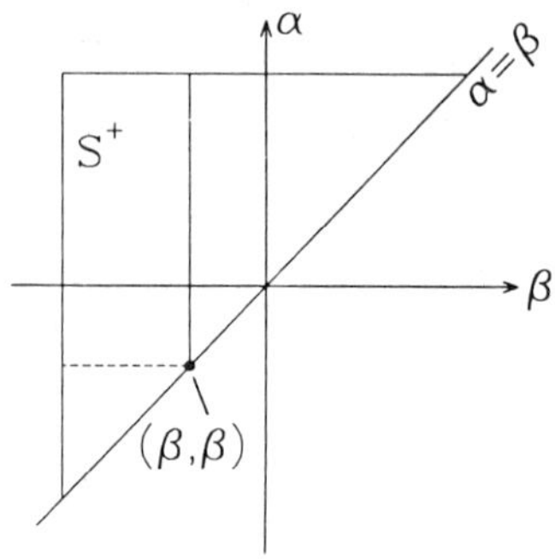

Fig. 2.5

it reaches some value β. The corresponding diagram is shown in Fig. 2.4. By using (2.12) and this diagram, we find that the output value $f_{\alpha\beta}$ corresponding to the final value of input is given by

$$f_{\alpha\beta} = \iint\limits_{R(\alpha,\beta_0,\beta,\beta)} \mu(\alpha',\beta')d\alpha'd\beta' - \iint\limits_{R(\alpha_0,\beta_0,\alpha,\beta)} \mu(\alpha',\beta')d\alpha'd\beta' + \frac{f_\beta^+ + f_\beta^-}{2}. \tag{2.14}$$

Here $R(\alpha_0, \beta_0, \alpha, \beta)$ is the rectangle whose opposite vertices are the points (α_0, β_0) and (α, β); the same meaning has the notation $R(\alpha, \beta_0, \beta, \beta)$.

Now suppose that, starting from the state of positive saturation, the input $u(t)$ is monotonically decreased until it reaches the value β. The corresponding $\alpha - \beta$ diagram is shown in Fig. 2.5. From this diagram and (2.12) we find the following expression for the resulting value of output:

$$f_\beta^- = \iint\limits_{R(\alpha_0,\beta_0,\beta,\beta)} \mu(\alpha',\beta')d\alpha'd\beta' + \frac{f_\beta^+ + f_\beta^-}{2}. \tag{2.15}$$

Next we introduce the function:

$$T(\alpha,\beta) = f_\beta^- - f_{\alpha\beta}, \tag{2.16}$$

which is equal to output increments between the limiting descending branch and first-order transition curves. From (2.14), (2.15), and (2.16) we find

$$\begin{aligned} T(\alpha,\beta) &= 2 \iint\limits_{R(\alpha_0,\beta_0,\alpha,\beta)} \mu(\alpha',\beta')d\alpha'd\beta' \\ &= 2\int_\alpha^{\alpha_0}\left(\int_{\beta_0}^{\beta} \mu(\alpha',\beta')d\beta\right)d\alpha. \end{aligned} \tag{2.17}$$

By differentiating the last expression twice, we obtain

$$\mu(\alpha,\beta) = -\frac{1}{2}\frac{\partial^2 T(\alpha,\beta)}{\partial\alpha\partial\beta}. \tag{2.18}$$

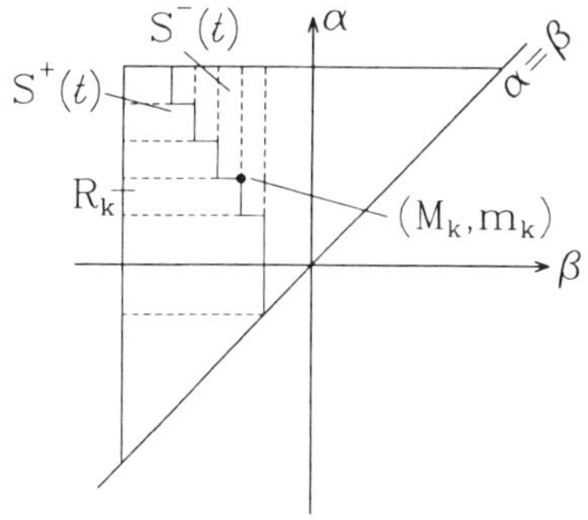

Fig. 2.6

By recalling the definition (2.16) of $T(\alpha, \beta)$, from (2.18) we derive

$$\mu(\alpha, \beta) = \frac{1}{2}\frac{\partial^2 f_{\alpha\beta}}{\partial\alpha\partial\beta}, \tag{2.19}$$

which coincides with the expression (1.29) obtained for the classical Preisach model.

In the conclusion of this section we show that the integration in (2.12) can be avoided and that the explicit expression for $f(t)$ in terms of experimentally measured function $T(\alpha, \beta)$ can be derived. To start the derivation, consider the $\alpha - \beta$ diagram shown in Fig. 2.6. Then according to (2.12) we have

$$\begin{aligned} f(t) &= \iint\limits_{S^+(t)} \mu(\alpha, \beta)d\alpha d\beta - \iint\limits_{S^-(t)} \mu(\alpha, \beta)d\alpha d\beta + \frac{f^+_{u(t)} + f^-_{u(t)}}{2} \\ &= 2\iint\limits_{S^+(t)} \mu(\alpha, \beta)d\alpha d\beta - \iint\limits_{R_{u(t)}} \mu(\alpha, \beta)d\alpha d\beta + \frac{f^+_{u(t)} + f^-_{u(t)}}{2}. \end{aligned} \tag{2.20}$$

To evaluate the integral over $R_{u(t)}$ in (2.20), we consider a monotonic increase of input from the state of negative saturation to some value $u(t)$. Then from (2.12) we find

$$f^+_{u(t)} = -\iint\limits_{R_{u(t)}} \mu(\alpha, \beta)d\alpha d\beta + \frac{f^+_{u(t)} + f^-_{u(t)}}{2}. \tag{2.21}$$

By substituting (2.21) into (2.20), we conclude

$$f(t) = 2\iint\limits_{S^+(t)} \mu(\alpha, \beta)d\alpha d\beta + f^+_{u(t)}. \tag{2.22}$$

It is clear that the moving model (2.12) is equivalent to the classical Preisach model as far as description of purely hysteretic behavior is concerned. For this reason, the wiping-out property and congruency property of minor loops are valid for the moving model. According to the wiping-out property, the staircase interface $L(t)$ in Fig. 2.6 has vertices whose α and β coordinates are equal to M_k and m_k, respectively, and, as before, numbers M_k and m_k form the alternating series of past dominant input extrema. Now we subdivide $S^+(t)$ into rectangles R_k and represent the integral in (2.22) as

$$\iint\limits_{S^+(t)} \mu(\alpha, \beta)d\alpha d\beta = \sum_{k=1}^{n(t)} \iint\limits_{R_k} \mu(\alpha, \beta)d\alpha d\beta. \tag{2.23}$$

It is easy to see from the expression (2.17) for the function $T(\alpha, \beta)$ that the integrals over R_k can be represented in the form

$$\iint\limits_{R_k} \mu(\alpha, \beta)d\alpha d\beta = T(M_{k+1}, m_k) - T(M_k, m_k). \tag{2.24}$$

From (2.22), (2.23), and (2.24), we obtain

$$f(t) = 2 \sum_{k=1}^{n(t)} [T(M_{k+1}, m_k) - T(M_k, m_k)] + f^+_{u(t)}. \tag{2.25}$$

This is the final result which expresses explicitly the output $f(t)$ in terms of experimentally measured function T.

2 Nonlinear Preisach Model of Hysteresis

The nonlinear (input dependent) Preisach model is discussed in this section. It has the following advantages over the classical model. First, the congruency property of minor loops is relaxed for the nonlinear model. This results in a broader area of applicability of this model as compared with the classical model. Second, the nonlinear model allows one to fit experimentally measured first- and second-order reversal curves. Since higher-order reversal curves are "sandwiched" between first- and second-order reversal curves, it is natural to expect that the nonlinear model will be more accurate than the classical one.

The nonlinear Preisach model of hysteresis can be mathematically defined as

$$f(t) = \iint\limits_{R_{u(t)}} \mu(\alpha, \beta, u(t))\hat{\gamma}_{\alpha\beta}u(t)d\alpha d\beta + \frac{f^+_{u(t)} + f^-_{u(t)}}{2}. \tag{2.26}$$

It is clear that a new feature of this model in comparison with the moving model (2.12) is the dependence of the distribution function μ on the current value of input $u(t)$. For this reason, the model (2.26) is termed as the nonlinear or input dependent Preisach model. Due to the new feature mentioned above,

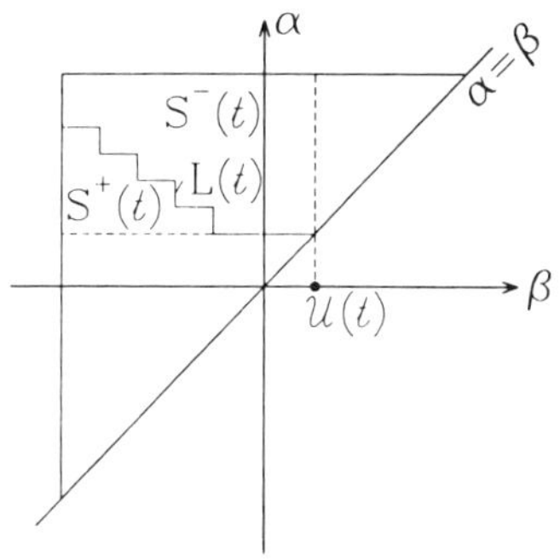

Fig. 2.7

the first term in the right-hand side of (2.26) can be construed as a partially reversible component of hysteresis nonlinearity. Indeed, for each pair (α, β), the integrand $\mu(\alpha, \beta, u(t))\hat{\gamma}_{\alpha\beta}u(t)$ is reversible for input variations between α and β.

The nonlinear Preisach model admits a geometric interpretation which is similar to that for the classical model. In other words, it could be easily demonstrated that at any instant of time t the rectangle $R_{u(t)}$ is subdivided into two sets (see Fig. 2.7): $S^+(t)$ and $S^-(t)$ consisting of points (α, β) for which $\hat{\gamma}_{\alpha\beta}u(t) = 1$ and $\hat{\gamma}_{\alpha\beta}u(t) = -1$, respectively. The interface $L(t)$ between $S^+(t)$ and $S^-(t)$ is a staircase line whose vertices have α and β coordinates equal (respectively) to past dominant extrema M_k and m_k. Using the above geometric interpretation, the model (2.26) can be represented in the form

$$f(t) = \iint_{S^+(t)} \mu(\alpha, \beta, u(t))\,d\alpha\,d\beta - \iint_{S^-(t)} \mu(\alpha, \beta, u(t))\,d\alpha\,d\beta + \frac{f^+_{u(t)} + f^-_{u(t)}}{2}. \qquad (2.27)$$

It can be shown that the following two properties are valid for the nonlinear model.

Property A (Wiping-Out Property)

Only the alternating series of past dominant extrema M_k and m_k are stored by the nonlinear Preisach model.

The proof of this property is completely identical to that for the classical model.

Property B (Property of Equal Vertical Chords)

All minor loops resulting from back-and-forth input variations between the same two consecutive extrema have equal vertical chords (output increments) for the same input values (Fig. 2.8).

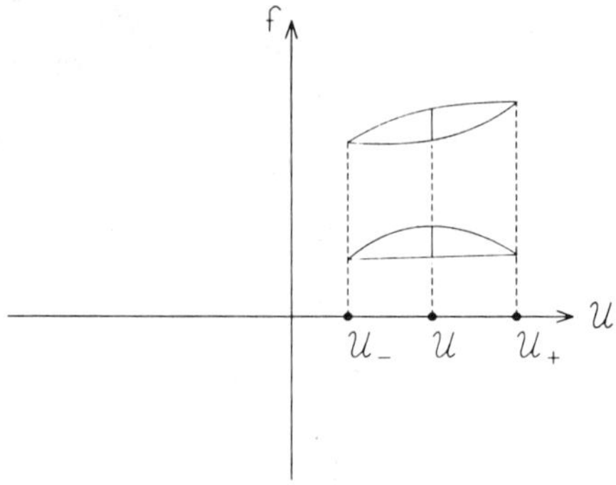

Fig. 2.8

Proof

Consider a minor loop formed as a result of back-and-forth input variations between u_+ and u_- (Fig. 2.9). Let f_u'' and f_u' be output values on descending and ascending branches of this loop, respectively. These values correspond to the same value of input u from the interval $u_- < u < u_+$. By using the expression (2.27), f_u'' and f_u' can be computed as

$$f_u' = \iint_{S'_+} \mu(\alpha, \beta, u) d\alpha d\beta - \iint_{S'_-} \mu(\alpha, \beta, u) d\alpha d\beta + \frac{f_u^+ + f_u^-}{2}, \tag{2.28}$$

$$f_u'' = \iint_{S''_+} \mu(\alpha, \beta, u) d\alpha d\beta - \iint_{S''_-} \mu(\alpha, \beta, u) d\alpha d\beta + \frac{f_u^+ + f_u^-}{2}, \tag{2.29}$$

where the sets S'_+, S'_-, S''_+, and S''_- are shown on the diagrams presented in Fig. 2.10. It is clear from these diagrams that

$$S''_+ - S'_+ = R(u_+, u, u_-), \qquad S'_- - S''_- = R(u_+, u, u_-). \tag{2.30}$$

From (2.28), (2.29), and (2.30), we derive

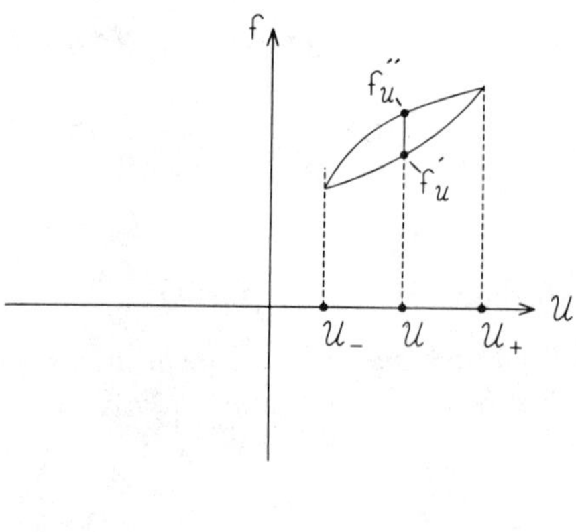

Fig. 2.9

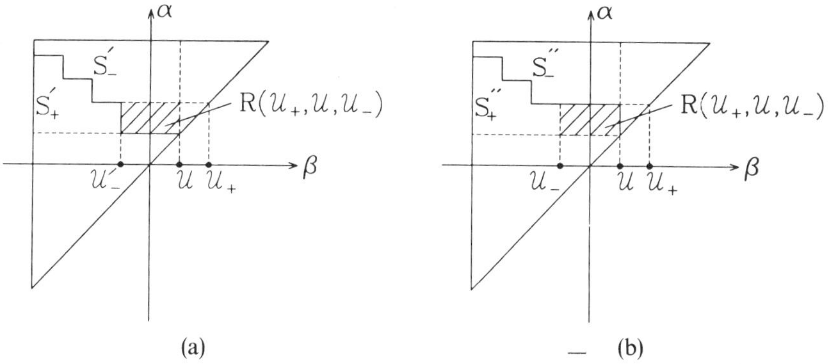

Fig. 2.10

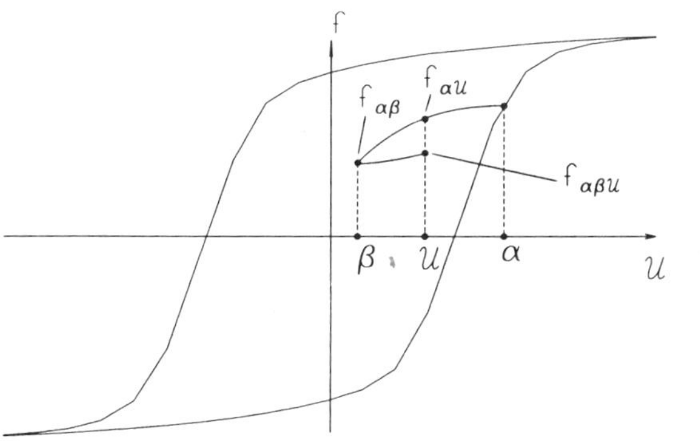

Fig. 2.11

$$f_u'' - f_u' = 2 \iint\limits_{R(u_+, u, u_-)} \mu(\alpha, \beta, u) d\alpha d\beta. \tag{2.31}$$

According to (2.31) we find that for any $u \in (u_+, u_-)$ the corresponding vertical chord does not depend on a particular past history preceding the formation of a minor loop. This proves that all comparable minor loops, that is, the loops with the same reversal input values u_+ and u_-, have equal vertical chords.

It is left to the reader as a useful exercise to prove that comparable minor loops described by the nonlinear model (2.26) are not necessarily congruent.

Now we turn to the identification problem of determining the distribution function $\mu(\alpha, \beta, u)$ by fitting the model (2.26) to some experimental data. It turns out that for the solution of this identification problem the sets of first- and second-order reversal curves are required. These curves can be measured experimentally as follows. We first decrease the input $u(t)$ to such a negative value that the outputs of all operators $\hat{\gamma}_{\alpha\beta}$ are equal to -1 (state of negative saturation). Then we monotonically increase the input until it reaches some value α. As we do this, we will follow along the ascending branch of the major loop (see Fig. 2.11). As we already know the first-order reversal curves are attached to this ascending branch and they are formed when the above monotonic increase of $u(t)$ is followed by a subsequent monotonic decrease. The notation $f_{\alpha u}^{+}$ will be used for the output values on the first-order reversal curve attached to the ascending branch of the major loop at the point f_α^+. The second-order reversal curves are attached to the first-order reversal curves, and they are formed when the above monotonic decrease is followed by a monotonic increase. The notation $f_{\alpha\beta u}$ will be used for the output values on the second-order reversal curve attached to the first-order reversal curve $f_{\alpha u}$ at the point $f_{\alpha\beta}$ (Fig. 2.11).

Consider the function

$$P(\alpha, \beta, u) = f_{\alpha u} - f_{\alpha\beta u}, \tag{2.32}$$

which has the physical meaning of output increments between the first- and second-order reversal curves.

It is clear from the definition of this function that

$$P(\alpha, u, u) = P(u, \beta, u) = P(u, u, u) = 0. \tag{2.33}$$

This property will be employed later in our discussion. Now we will try to relate the function $P(\alpha, \beta, u)$ to the distribution function $\mu(\alpha, \beta, u)$. To this end, we will use the diagrams shown in Fig. 2.12. From these diagrams and (2.27), we conclude

$$\begin{aligned} f_{\alpha u} &= \iint_{\tilde{S}_+} \mu(\alpha', \beta', u)\, d\alpha' d\beta' - \iint_{\tilde{S}_-} \mu(\alpha', \beta', u)\, d\alpha' d\beta' + \frac{f_u^+ + f_u^-}{2}, \\ f_{\alpha\beta u} &= \iint_{\tilde{\tilde{S}}_+} \mu(\alpha', \beta', u)\, d\alpha' d\beta' - \iint_{\tilde{\tilde{S}}_-} \mu(\alpha', \beta', u)\, d\alpha' d\beta' + \frac{f_u^+ + f_u^-}{2}, \end{aligned} \tag{2.34}$$

$$\tilde{S}_+ - \tilde{\tilde{S}}_+ = R(\alpha, \beta, u), \qquad \tilde{\tilde{S}}_- - \tilde{S}_- = R(\alpha, \beta, u). \tag{2.35}$$

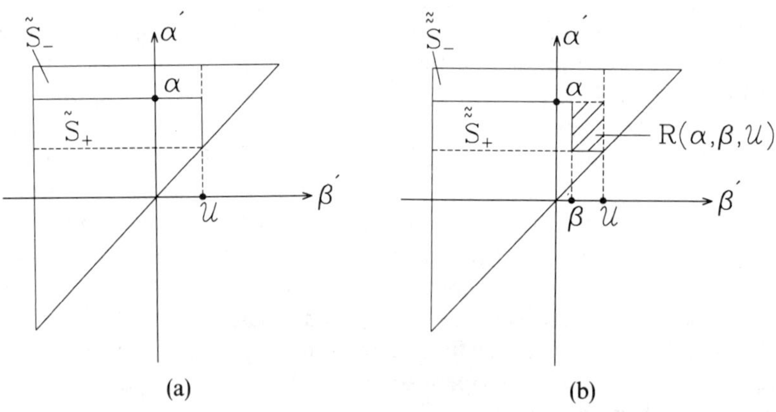

Fig. 2.12

By using (2.34) and (2.35), we derive

$$P(\alpha, \beta, u) = 2 \iint_{R(\alpha, \beta, u)} \mu(\alpha', \beta', u) d\alpha' d\beta' = 2 \int_u^\alpha \left(\int_\beta^u \mu(\alpha', \beta', u) d\beta' \right) d\alpha'. \quad (2.36)$$

By differentiating the last expression twice, we obtain

$$\mu(\alpha, \beta, u) = -\frac{1}{2} \frac{\partial^2 P(\alpha, \beta, u)}{\partial\alpha \partial\beta}. \quad (2.37)$$

By recalling the definition (2.32) of $P(\alpha, \beta, u)$, from (2.37) we find

$$\mu(\alpha, \beta, u) = \frac{1}{2} \frac{\partial^2 f_{\alpha\beta u}}{\partial\alpha \partial\beta}. \quad (2.38)$$

It is clear from the above derivation that the expressions (2.37) and (2.38) are valid if $\beta < u < \alpha$. If $u \leq \beta$ or $u \geq \alpha$, then we define

$$\mu(\alpha, \beta, u) \equiv 0, \quad (2.39)$$

which is consistent with (2.33).

Thus, if the distribution function $\mu(\alpha, \beta, u)$ is determined from (2.36) or (2.38), then the nonlinear model (2.26) will match the increments between first- and second-order reversal curves. Next we shall show that the limiting ascending branch will be matched by this model as well. Indeed, it is easy to see that the output values f_u on the limiting ascending branch predicted by the nonlinear model (2.26) are equal to

$$f_u = - \iint_{R(\alpha_0, \beta_0, u)} \mu(\alpha', \beta', u) d\alpha' d\beta' + \frac{f_u^+ + f_u^-}{2}. \quad (2.40)$$

According (2.36) and (2.32), we have

$$\iint_{R(\alpha_0, \beta_0, u)} \mu(\alpha', \beta', u) d\alpha d\beta' = \frac{1}{2} P(\alpha_0, \beta_0, u) = \frac{1}{2}(f_{\alpha_0 u} - f_{\alpha_0 \beta_0 u})$$

$$= \frac{1}{2}(f_u^- - f_u^+). \quad (2.41)$$

By substituting (2.41) into (2.40), we find

$$f_u = f_u^+.$$

Thus, if the function, $\mu(\alpha, \beta, u)$, is determined according to (2.36) or (2.38), then the nonlinear model (2.26) fits (a) the output increments between the first- and second-order reversal curves and (b) the ascending branch of the major loop. Since the ascending branch of the major loop can be construed as a second-order reversal curve, we conclude that the nonlinear model (2.26) fits the sets of first- and second-order reversal curves.

In the above discussion we have used second-order *increasing* reversal curves (see Fig. 2.13) in order to determine the distribution function, $\mu(\alpha, \beta, u)$. However, by almost literally repeating the previous line of reasoning, a similar expression can be found for $\mu(\alpha, \beta, u)$ by using second-order *decreasing* reversal curves $f_{\tilde{\beta}\tilde{\alpha}\tilde{u}}$. One of these curves is shown in Fig. 2.14. By using this figure, we can introduce the function

$$P(\tilde{\beta}, \tilde{\alpha}, \tilde{u}) = f_{\tilde{\beta}\tilde{\alpha}\tilde{u}} - f_{\tilde{\beta}\tilde{u}}. \tag{2.42}$$

In the same way as before, we can show that

$$P(\tilde{\beta}, \tilde{\alpha}, \tilde{u}) = 2 \iint_{R(\tilde{\alpha}, \tilde{\beta}, \tilde{u})} \mu(\alpha', \beta', \tilde{u})\, d\alpha'\, d\beta' \tag{2.43}$$

and

$$\mu(\tilde{\alpha}, \tilde{\beta}, \tilde{u}) = -\frac{1}{2} \frac{\partial^2 P(\tilde{\beta}, \tilde{\alpha}, \tilde{u})}{\partial \tilde{\alpha} \partial \tilde{\beta}}. \tag{2.44}$$

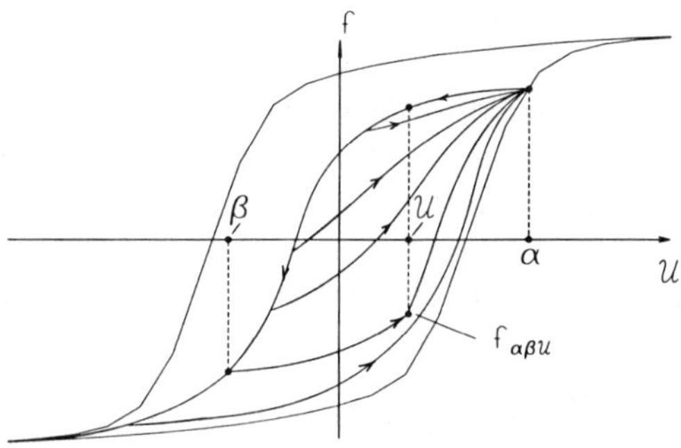

Fig. 2.13

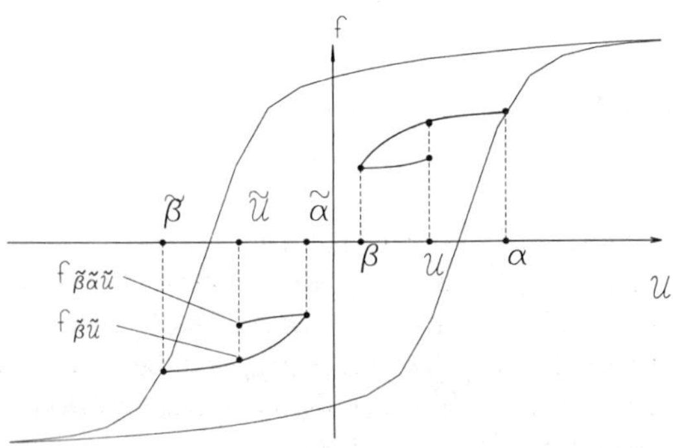

Fig. 2.14

If

$$\tilde{\beta} = -\alpha, \qquad \tilde{\alpha} = -\beta, \qquad \tilde{u} = -u, \tag{2.45}$$

then due to the symmetry between increasing and decreasing second-order reversal curves (see Fig. 2.14), we find

$$P(\tilde{\beta}, \tilde{\alpha}, \tilde{u}) = P(\alpha, \beta, u). \tag{2.46}$$

By using (2.45) and (2.46) in (2.44), we derive

$$\mu(-\beta, -\alpha, -u) = -\frac{1}{2}\frac{\partial^2 P(\alpha, \beta, u)}{\partial\alpha\partial\beta}. \tag{2.47}$$

From (2.47) and (2.37), we conclude

$$\mu(\alpha, \beta, u) = \mu(-\beta, -\alpha, -u). \tag{2.48}$$

The last formula can be regarded as a generalization of the mirror symmetry (1.46) previously established for the classical Preisach model.

We next proceed to the proof of the following important result.

Representation Theorem

The wiping-out property and the property of equal vertical chords for minor loops constitute the necessary and sufficient conditions for the representation of a hysteresis nonlinearity by the nonlinear Preisach model on the set of piecewise monotonic inputs.

Proof

Necessity

If a hysteresis nonlinearity is representable by the nonlinear Preisach model, then this nonlinearity should have the same properties as the model. This means that this nonlinearity should exhibit the wiping-out property and the property of equal vertical chords for comparable minor loops.

Sufficiency

Consider a hysteresis nonlinearity which has both the wiping-out property and the property of equal vertical chords. For this nonlinearity we find the distribution function $\mu(\alpha, \beta, u)$ by using formula (2.36) (or (2.38)). Then the nonlinear Preisach model (2.26) will match exactly the sets of all first- and second-order reversal curves. We intend to prove that this model will match all possible higher-order reversal curves as well. The proof is based on the induction argument. Let us assume that the above statement is true for all possible reversal curves up to the order $k - 1$. Then, for the induction inference to take place, we need to prove that the same statement holds for any reversal curve of order k. Let a be a point at which a reversal curve of order k is attached to a reversal curve of order $k - 1$ (see Fig. 2.15). According to the induction assumption, the output values for the actual hysteresis nonlinearity and for

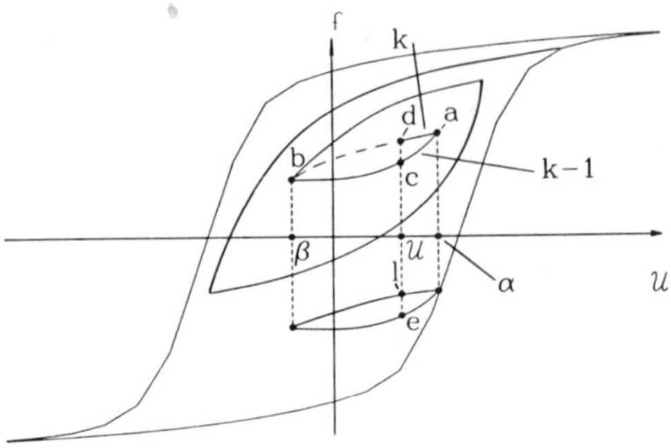

Fig. 2.15

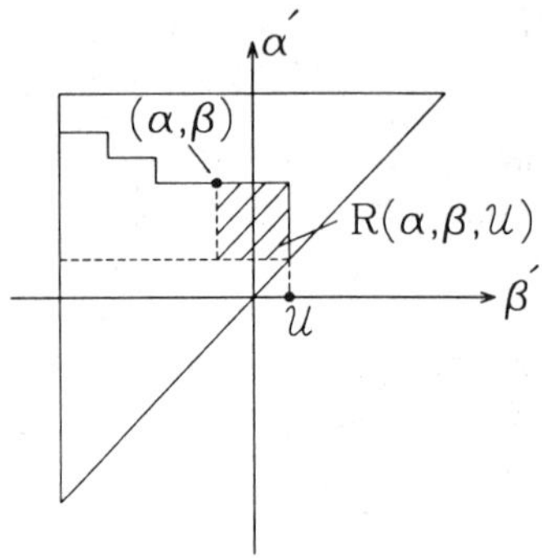

Fig. 2.16

the nonlinear Preisach model coincide at each point of the reversal curve of the order $k-1$. Thus, it remains to be proven that the output increments between the reversal curves of orders k and $k-1$ are the same for the actual hysteresis nonlinearity and for the nonlinear model (2.26). It is here that the wiping-out property and the property of equal vertical chords will be used. According to the wiping-out property, the k-th reversal curve should meet the $k-1$th reversal curve at the point b, which is the point of inception of the latter curve. As a result, a minor loop is formed. Consider a comparable minor loop (loop with the same reversal input values), which is attached to the limiting ascending branch (Fig. 2.15). This loop is formed by some first- and second-order transition curves. According to the property of equal vertical chords, this loop has the same vertical chords as the loop formed by $k-1$th and kth reversal curves. Consider an arbitrary value u of input such that $\beta < u < \alpha$. Using the diagram shown in Fig. 2.16, it is easy to derive the following expression for the output increment

$$f_u^{(k)} - f_u^{(k-1)} = 2 \iint_{R(\alpha,\beta,u)} \mu(\alpha', \beta', u)\,d\alpha'\,d\beta', \tag{2.49}$$

where $f_u^{(k)}$ and $f_u^{(k-1)}$ are the output values on the k-th and $k-1$th reversal curves, respectively, corresponding to the input value u and predicted by the model (2.26). From (2.36) we find that the right-hand side of (2.49) is equal to the output increment el between the first- and second-order transition curves. Consequently,

$$f_u^{(k)} - f_u^{(k-1)} = el. \tag{2.50}$$

According to the property of equal vertical chords we have

$$el = cd. \tag{2.51}$$

From (2.50) and (2.51), we conclude

$$f_u^{(k)} - f_u^{(k-1)} = cd. \tag{2.52}$$

Thus, the nonlinear model predicts the correct output increments between the k-th and $(k-1)$th reversal curves. From this fact and the above induction assumption we conclude that the nonlinear model predicts the correct output values on the reversal curve of order k. This concludes the proof of the theorem.

The proven theorem establishes the exact bounds of applicability of the nonlinear Preisach model. It is apparent that the property of equal vertical chords is more general than the congruency property. Indeed, if comparable minor loops are congruent, then they have equal vertical chords. However, if comparable minor loops have equal vertical chords, they are not necessarily congruent. This clearly shows that the nonlinear Preisach model (2.26) has a broader area of applicability than the classical Preisach model. Next we shall show that the nonlinear model (2.26) contains the classical Preisach model as a particular case. The exact statement of this fact is given by the following theorem.

Reduction Theorem

If all comparable minor loops of hysteresis nonlinearity are congruent, then the nonlinear Preisach model for this nonlinearity coincides with the classical Preisach model as far as purely hysteretic behavior of this nonlinearity is concerned.

Proof

Since all comparable minor loops are congruent, these loops have equal vertical chords. Assuming also that the wiping-out property holds, we can represent the mentioned hysteresis nonlinearity by the nonlinear Preisach model (2.26). We next show that, because of the congruency property, the distribution function μ in (2.26) does not depend on u. It is clear from the Fig. 2.17 that the congruency property results in the congruency of second-order reversal curves $f_{\alpha'\beta u}$, $f_{\alpha\beta u}$, and $f_{\alpha''\beta u}$. This means that the derivative $\partial f_{\alpha\beta u}/\partial u$

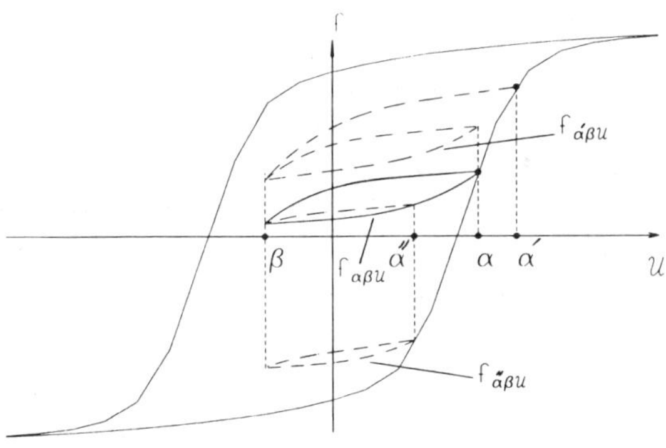

Fig. 2.17

does not depend on α. Consequently,

$$\frac{\partial^3 f_{\alpha\beta u}}{\partial\alpha\partial\beta\partial u} \equiv 0. \tag{2.53}$$

From (2.53) and (2.38), we find

$$\frac{\partial \mu(\alpha, \beta, u)}{\partial u} \equiv 0. \tag{2.54}$$

Thus, μ does not depend on u and is only a function of α and β:

$$\mu(\alpha, \beta, u) = \nu(\alpha, \beta). \tag{2.55}$$

By substituting (2.55) into (2.26), we obtain

$$f(t) = \iint_{R_{u(t)}} \nu(\alpha, \beta)\hat{\gamma}_{\alpha\beta}u(t)\,d\alpha\,d\beta + \frac{f^+_{u(t)} + f^-_{u(t)}}{2}. \tag{2.56}$$

It remains to be proven that $\nu(\alpha, \beta)$ in (2.56) coincides with $\mu(\alpha, \beta)$ in (2.12). The proof is straightforward. By using (2.12) and (2.56), we find that for any α, β and u such that $\beta \le u \le \alpha$ we have

$$P(\alpha, \beta, u) = f_{\alpha u} - f_{\alpha\beta u} = 2 \iint_{R(\alpha,\beta,u)} \mu(\alpha', \beta')\,d\alpha'\,d\beta', \tag{2.57}$$

$$P(\alpha, \beta, u) = f_{\alpha u} - f_{\alpha\beta u} = 2 \iint_{R(\alpha,\beta,u)} \nu(\alpha', \beta')\,d\alpha'\,d\beta'. \tag{2.58}$$

From (2.57) and (2.58), as before, we derive

$$\mu(\alpha, \beta) = \nu(\alpha, \beta) = -\frac{1}{2}\frac{\partial^2 P(\alpha, \beta, u)}{\partial\alpha\partial\beta}. \tag{2.59}$$

Thus, it is proven that under the congruency condition, the nonlinear Preisach model (2.26) coincides with the moving Preisach model (2.12). On the other hand, the moving model coincides with the classical Preisach model as far as the description of purely hysteretic behavior is concerned. From here we conclude that under the congruency condition the nonlinear model (2.26) coincides with the classical Preisach model, which concludes the proof of the theorem.

We next turn to the discussion of numerical implementation of the nonlinear model (2.26). We intend to show that double integration in (2.26) can be completely avoided and that explicit expressions for the output $f(t)$ in terms of experimentally measured function $P(\alpha, \beta, u)$ can be derived. The starting point of our derivation is the expression (2.27) which can be modified as follows:

$$f(t) = \iint_{R_{u(t)}} \mu(\alpha, \beta, u(t))\,d\alpha\,d\beta - 2\iint_{S^-(t)} \mu(\alpha, \beta, u(t))\,d\alpha\,d\beta + \frac{f^+_{u(t)} + f^-_{u(t)}}{2}. \tag{2.60}$$

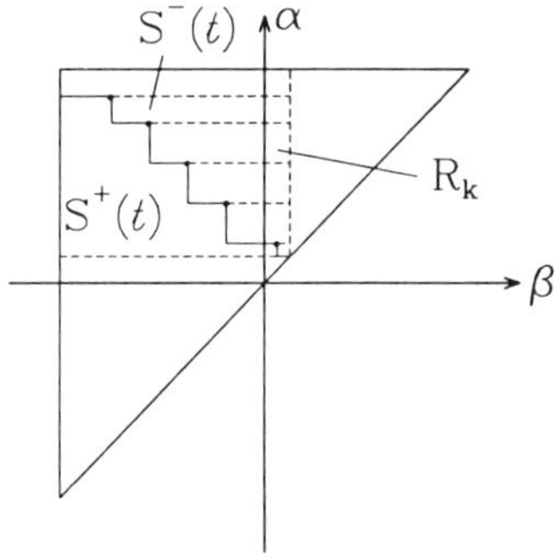

Fig. 2.18

According to (2.41), we have

$$\iint_{R_{u(t)}} \mu(\alpha, \beta, u(t))d\alpha d\beta = \frac{f^-_{u(t)} - f^+_{u(t)}}{2}. \tag{2.61}$$

By substituting (2.61) into (2.60), we obtain

$$f(t) = f^-_{u(t)} - 2 \iint_{S^-(t)} \mu(\alpha, \beta, u(t))d\alpha d\beta. \tag{2.62}$$

We next subdivide $S^-(t)$ into rectangles R_k (see Fig. 2.18) and represent the integral in (2.62) as

$$\iint_{S^-(t)} \mu(\alpha, \beta, u(t))d\alpha d\beta = \sum_{k=1}^{n(t)} \iint_{R_k} \mu(\alpha, \beta, u(t))d\alpha d\beta. \tag{2.63}$$

It is easy to see that

$$R_k = R(M_k, m_k, u(t)) - R(M_{k+1}, m_k, u(t)). \tag{2.64}$$

From (2.64) and (2.36), we conclude

$$\begin{aligned}\iint_{R_k} \mu(\alpha, \beta, u(t))d\alpha d\beta &= \iint_{R(M_k, m_k, u(t))} \mu(\alpha, \beta, u(t))d\alpha d\beta \\ &\quad - \iint_{R(M_{k+1}, m_k, u(t))} \mu(\alpha, \beta, u(t))d\alpha d\beta \\ &= \frac{1}{2}[P(M_k, m_k, u(t)) - P(M_{k+1}, m_k, u(t))]. \end{aligned} \tag{2.65}$$

By substituting (2.65) into (2.63) and then (2.63) into (2.62), we obtain

$$f(t) = f^-_{u(t)} + \sum_{k=1}^{n(t)} [P(M_{k+1}, m_k, u(t)) - P(M_k, m_k, u(t))]. \tag{2.66}$$

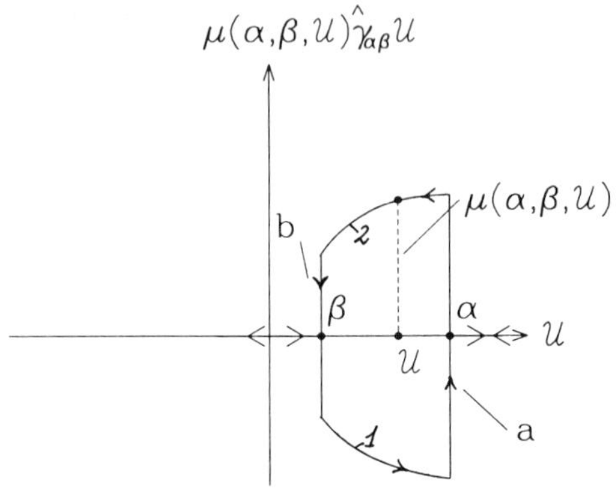

Fig. 2.19

This is the final formula which expresses explicitly the output $f(t)$ in terms of the experimentally measured function P. This formula has been used to develop a digital code which numerically implements the nonlinear Preisach model. This code computes output values by using input values, a set of second-order reversals curves, and an input history which are all specified by the user. The structure of the code is very similar to the structure of the code which implements the classical Preisach model. For this reason, the detailed discussion of this code is omitted.

We next proceed to the discussion of hysteretic energy losses for hysteresis nonlinearities described by the nonlinear Preisach model (2.26). We begin with the case of elementary hysteresis nonlinearity represented by the expression $\mu(\alpha, \beta, u(t))\hat{\gamma}_{\alpha\beta}u(t)$. It follows from (2.39) that the input-output relationship for this nonlinearity is described by a loop shown in Fig. 2.19. It is clear that the branch 1 of this loop is fully reversible except for its part a along which the irreversible switching up occurs. Similarly, the branch 2 of the loop is fully reversible except for its part b along which the irreversible switching down occurs. If a periodic variation of input is such that the whole loop is traced, then the hysteretic energy loss for one cycle, Q_{cycle}, is equal to the area enclosed by the loop

$$Q_{\text{cycle}} = 2 \int_{\beta}^{\alpha} \mu(\alpha, \beta, u(t))du. \tag{2.67}$$

It is clear that this energy loss occurs as a result of irreversible switchings along the parts a and b of the above loop. Consequently,

$$Q_{\text{cycle}} = q_{\alpha} + q_{\beta}, \tag{2.68}$$

where q_{α} and q_{β} are losses occurred as a result of irreversible "up" and "down" switchings, respectively. Due to the symmetry between branches 1 and 2, it can be concluded that

$$q_{\alpha} = q_{\beta}. \tag{2.69}$$

From (2.67), (2.68), and (2.69) we find

$$q_\alpha = q_\beta = \int_\beta^\alpha \mu(\alpha, \beta, u)du. \tag{2.70}$$

In the nonlinear Preisach model, any input variation is associated with irreversible switchings of some elementary loops $\mu(\alpha, \beta, u(t))\hat{\gamma}_{\alpha\beta}u(t)$. It is natural to equate the hysteretic energy loss occurring for some input variation to the sum of energy losses resulting from the irreversible switchings of the above elementary loops. Since in the nonlinear Preisach model we deal with continuous ensembles of the elementary hysteresis loops, the above summation should be replaced by integration. Thus, if Ω denotes the region of points on the $\alpha - \beta$ diagram for which the elementary hysteresis loops are irreversibly switched during some monotonic input variation, then the hysteretic loss Q for this input variation is given by

$$Q = \iint_\Omega \left(\int_\beta^\alpha \mu(\alpha, \beta, u)du \right) d\alpha d\beta. \tag{2.71}$$

In a particular case when the distribution function μ does not depend on u, from (2.71) we obtain the expression

$$Q = \iint_\Omega \mu(\alpha, \beta)(\alpha - \beta)d\alpha d\beta, \tag{2.72}$$

which has been previously derived for the classical Preisach model (see Chapter I).

A typical shape of the region Ω is shown in Fig. 2.20. The region Ω can be subdivided into a triangle and some trapezoids which in turn can be represented as differences of triangles. Thus, if the integral (2.71) can be evaluated for any triangle region, then it will be easy to determine this integral for any possible shape of Ω. In the case when Ω is a triangle, a very simple expression for the integral (2.71) in terms of experimentally measured function P can be derived. The derivation proceeds as follows. Let $T(u_+, u_-)$ be a triangle (see

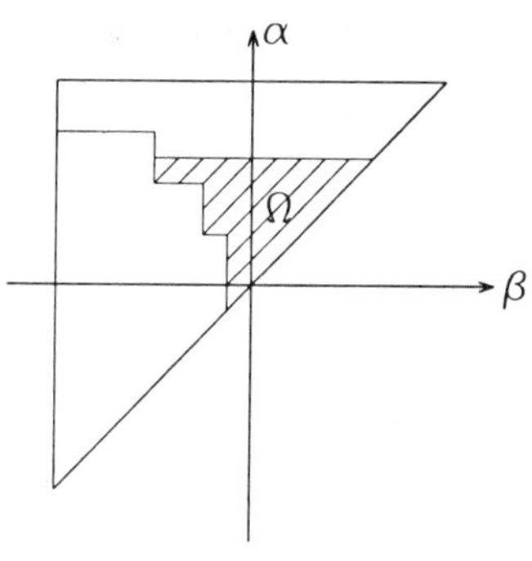

Fig. 2.20

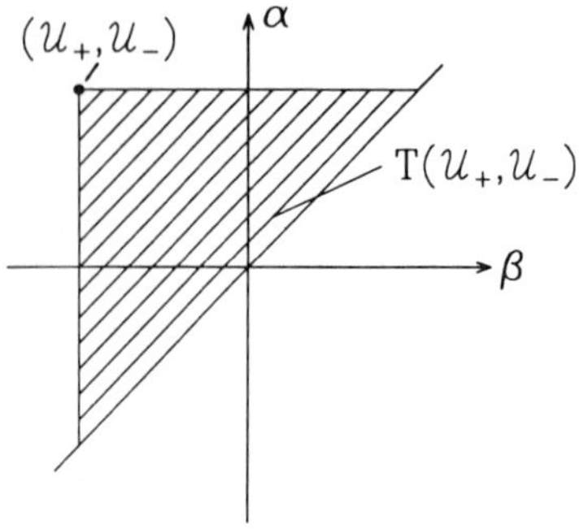

Fig. 2.21

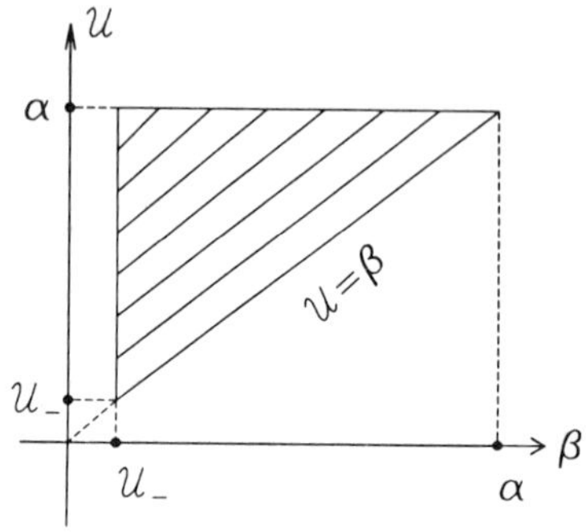

Fig. 2.22

Fig. 2.21) swept during a monotonic increase of input from some local minimum u_- to some value u_+. According to (2.71), this input variation results in the hysteretic loss:

$$Q(u_-, u_+) = \iint\limits_{T(u_+,u_-)} \left(\int_\beta^\alpha \mu(\alpha, \beta, u \right) du) d\alpha d\beta$$

$$= \int_{u_-}^{u_+} \left(\int_{u_-}^{\alpha} \left(\int_\beta^\alpha \mu(\alpha, \beta, u) du \right) d\beta \right) d\alpha. \tag{2.73}$$

By using Fig. 2.22, we can change the order of integration in the two interior integrals as follows:

$$\int_{u_-}^{\alpha} \left(\int_\beta^\alpha \mu(\alpha, \beta, u) du \right) d\beta = \int_{u_-}^{\alpha} \left(\int_{u_-}^{u} \mu(\alpha, \beta, u) d\beta \right) du. \tag{2.74}$$

Now, by invoking the formula (2.37), from (2.74) we find

$$\int_{u_-}^{\alpha} \left(\int_\beta^\alpha \mu(\alpha, \beta, u) du \right) d\beta = -\frac{1}{2} \int_{u_-}^{\alpha} \left(\int_{u_-}^{u} \frac{\partial^2 P(\alpha, \beta, u)}{\partial\alpha\partial\beta} d\beta \right) du$$

$$= -\frac{1}{2} \int_{u_-}^{\alpha} \left[\frac{\partial P(\alpha, u, u)}{\partial\alpha} - \frac{\partial P(\alpha, u_-, u)}{\partial\alpha} \right] du. \tag{2.75}$$

By substituting (2.75) into (2.73) and by changing this time the order of integration according to Fig. 2.23, we derive:

$$Q(u_-, u_+) = -\frac{1}{2} \int_{u_-}^{u_+} \left(\int_{u_-}^{\alpha} \left[\frac{\partial P(\alpha, u, u)}{\partial\alpha} - \frac{\partial P(\alpha, u_-, u)}{\partial\alpha} \right] du \right) d\alpha$$

$$= -\frac{1}{2} \int_{u_-}^{u_+} \left(\int_{u}^{u_+} \left[\frac{\partial P(\alpha, u, u)}{\partial\alpha} - \frac{\partial P(\alpha, u_-, u)}{\partial\alpha} \right] d\alpha \right) du$$

$$= -\frac{1}{2} \int_{u_-}^{u_+} [P(u_+, u, u) - P(u, u, u) - P(u_+, u_-, u)$$

$$+ P(u, u_-, u)] du. \tag{2.76}$$

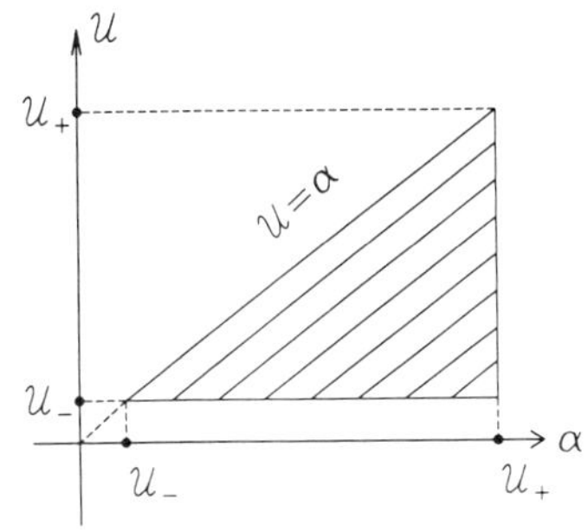

Fig. 2.23

By using (2.33) in (2.76), we finally obtain

$$Q(u_-, u_+) = \frac{1}{2}\int_{u_-}^{u_+} P(u_+, u_-, u)du. \tag{2.77}$$

The formula (2.77) is remarkably simple if compared with a similar formula (1.115) derived for the classical Preisach model. The formula (2.77) allows one to evaluate hysteretic energy losses for arbitrary input variations. Indeed, consider some input variation for which the region Ω has the shape shown in Fig. 2.20. Then, the corresponding hysteretic energy loss is given by (2.71) which can be written as

$$Q = Q(M_n, m_n) + \sum_{i=1}^{k} [Q(M_{n-i}, m_{n-i}) - Q(M_{n-i}, m_{n-i-1})]. \tag{2.78}$$

Each term in (2.78) can be evaluated by employing the formula (2.77).

The formula (2.77) admits the following geometric interpretation. Consider a minor loop formed as a result of back-and-forth input variations between two consecutive extrema u_- and u_+. If a hysteresis nonlinearity is representable by the nonlinear Preisach model, then all comparable minor loops of this nonlinearity have equal vertical chords. Consequently, $P(u_+, u_-, u)$ can be construed as a vertical chord of the above minor loop. This implies that the integral in (2.77) is equal to the area A of this loop. Thus, the expression (2.77) can be written as

$$Q(u_-, u_+) = \frac{A}{2}. \tag{2.79}$$

From (2.79) we immediately infer the following result.

Theorem

For any loop of hysteresis nonlinearity representable by the nonlinear Preisach model, hysteretic losses occurring along ascending and descending branches are the same.

By using this theorem we can relate hysteretic losses occurring for arbitrary input variations to certain cyclic losses. This can be accomplished in exactly

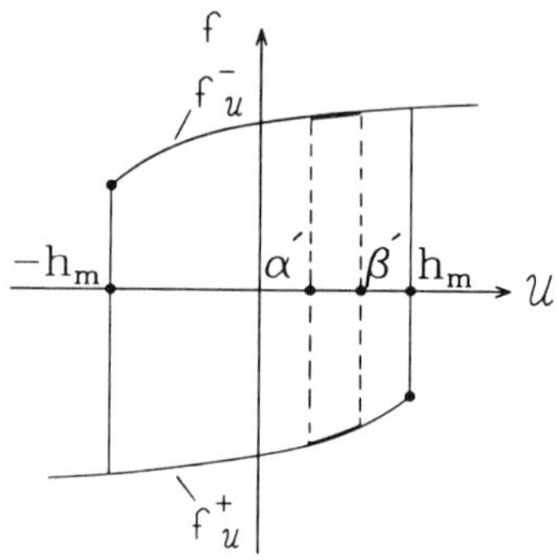

Fig. 2.24

the same way as it was done for the classical Preisach model. In particular, the expressions (1.122) and (1.123) can be derived.

We conclude this section by the discussion of two facts which may facilitate further appreciation of the advantages of the nonlinear Preisach model over the classical one. Consider the hysteresis nonlinearities exhibited by the Stoner-Wohlfarth magnetic particles (see Fig. 2.24). It is pointed out in Chapter I that these nonlinearities can not be described by the classical Preisach model. The reason is that degenerated minor loops traced as the input varies back and forth between α' and β' are not congruent. It is easy to see that the hysteresis nonlinearities exhibited by the Stoner-Wohlfarth magnetic particles can be described by the nonlinear Preisach model. First, this is clear from the fact that the property of equal vertical chords is satisfied because all minor loops are degenerated, and, consequently, they have equal (zero) vertical chords. Second, the explicit representation (2.26) for the Stoner-Wohlfarth hysteresis nonlinearities can be found. Indeed, it can be shown that

$$\mu(\alpha, \beta, u(t)) = \frac{f_{u(t)}^- - f_{u(t)}^+}{2} \delta(\alpha - h_m, \beta + h_m), \tag{2.80}$$

where δ is the Dirac function.

By substituting (2.80) into (2.26), we find

$$f(t) = \frac{f_{u(t)}^- - f_{u(t)}^+}{2} \hat{\gamma}_{h_m, -h_m} u(t) + \frac{f_{u(t)}^- + f_{u(t)}^+}{2}. \tag{2.81}$$

It can be easily checked that the expression (2.81) indeed represents the hysteresis nonlinearities depicted in Fig. 2.24.

Finally, the following observation may help the reader appreciate the extent to which the nonlinear Preisach model is more general than the classical one. The classical Preisach model represents hysteresis nonlinearities which for any *reversal* point have (regardless of past history) only one branch starting from this point (see Fig. 2.25a). This branch is congruent to one of the first-order reversal curves. This is an apparent consequence of the congruency property

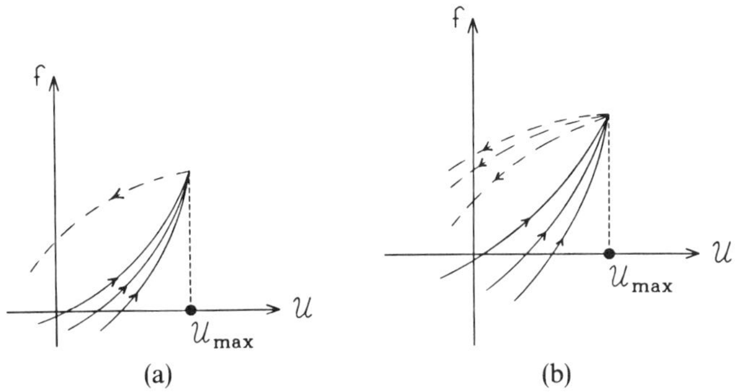

Fig. 2.25

of the classical model. On the other hand, the nonlinear Preisach model (2.26) describes hysteresis nonlinearities with the property that for any reversal point there are an infinite number of possible branches starting at this point (see Fig. 2.25b). A particular realization of these branches is determined by a particular past history. This property follows, for instance, from the expression (2.66) where all terms are input-dependent. For the classical model, the similar expressions (1.71) and (1.74) have only the last terms which are input-dependent. The above observation shows that the nonlinear model is endowed by a much more general mechanism of branching than the classical model.

3 Restricted Preisach Models of Hysteresis

In the previous section we have discussed the nonlinear Preisach model of hysteresis which is a far-reaching generalization of the classical Preisach model. This generalization has been achieved by assuming that the distribution function μ is dependent of the current value of input $u(t)$. Another approach to the generalization of the classical Preisach model is to assume that the function μ depends on stored past extremum values of input, $\{M_k, m_k\}$. This approach has been recently and briefly explored in [45]. Here, we shall follow this approach as well, however, our treatment of the model itself and the identification problem for this model will deviate appreciably from the discussion in [45].

We begin with the simplest case when the function μ depends only on the first global maximum M_1. Thus, we define new Preisach-type model as

$$f(t) = \iint_{T_{M_1}} \mu(\alpha, \beta, M_1)\hat{\gamma}_{\alpha\beta}u(t)d\alpha d\beta + C_{M_1}, \tag{2.82}$$

where M_1 is the largest input maximum since the departure from the state of negative saturation, and the support of $\mu(\alpha, \beta, M_1)$ is the triangle T_{M_1} defined

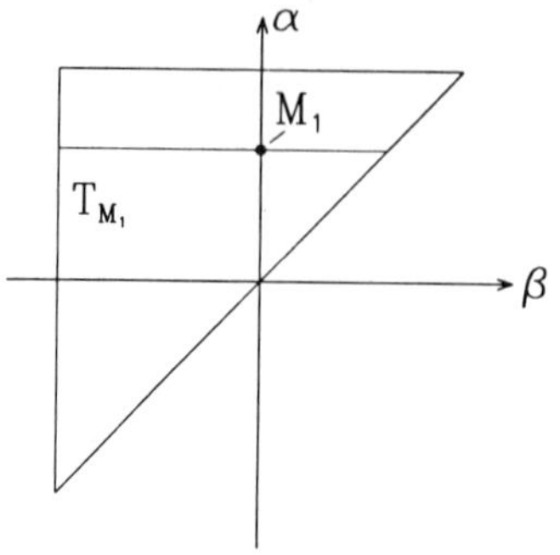

Fig. 2.26

by inequalities $\beta_0 \le \beta \le \alpha \le M_1$ (see Fig. 2.26). We shall use the following interpretation of the above model. For any fixed extremum M_1, the model (2.82) will be used to describe hysteresis behavior in the region confined between an ascending branch $f^+_{u(t)}$ and a first-order reversal curve $f_{M_1\beta}$ (see Fig. 2.27). For this reason, the model (2.82) can be termed as a "restricted" Preisach model of hysteresis. However, the above "restriction" does not diminish at all the region of applicability of the model. This is because M_1 may assume any value between β_0 and α_0.

The constant C_{M_1} in (2.82) can be easily determined from the condition that the model matches the output values $f^+_{M_1}$ for $u(t) = M_1$ and f_- for $u(t) = \beta_0$. Indeed, from (2.82) we easily derive

$$f^+_{M_1} = \iint_{T_{M_1}} \mu(\alpha, \beta, M_1)d\alpha d\beta + C_{M_1}, \tag{2.83}$$

$$f_- = -\iint_{T_{M_1}} \mu(\alpha, \beta, M_1)d\alpha d\beta + C_{M_1}. \tag{2.84}$$

From (2.83) and (2.84), we find

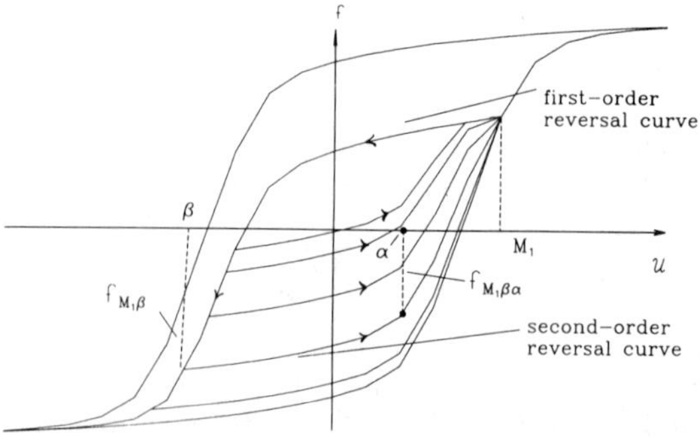

Fig. 2.27

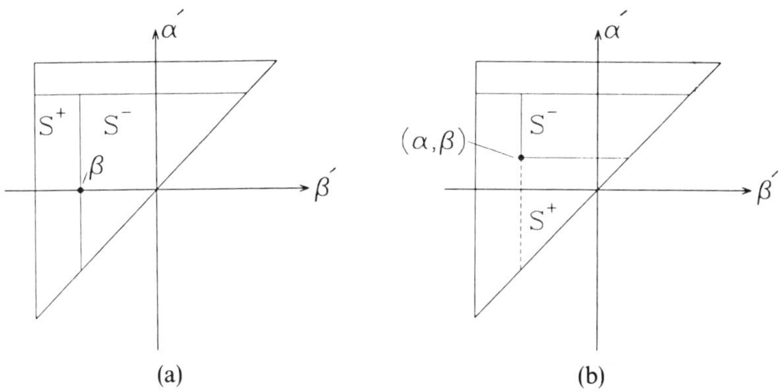

Fig. 2.28

$$C_{M_1} = \frac{f_{M_1}^+ + f_-}{2}, \tag{2.85}$$

which leads to the following representation for the model (2.82):

$$f(t) = \iint\limits_{T_{M_1}} \mu(\alpha, \beta, M_1)\hat{\gamma}_{\alpha\beta}u(t)d\alpha d\beta + \frac{f_{M_1}^+ + f_-}{2}. \tag{2.86}$$

Next we shall be concerned with the solution of the identification problem. The essence of this problem is in determining the function $\mu(\alpha, \beta, M_1)$ by fitting the model (2.86) to some experimental data. It turns out that $\mu(\alpha, \beta, M_1)$ can be found by matching second-order reversal curves $f_{M_1\beta\alpha}$ shown in Fig. 2.27. To do this, we introduce the function

$$F(\alpha, \beta, M_1) = \tfrac{1}{2}(f_{M_1\beta\alpha} - f_{M_1\beta}). \tag{2.87}$$

This function is equal to one-half of the output increments along the second-order reversal curves. To relate this function to $\mu(\alpha, \beta, M_1)$, consider diagrams shown in Fig. 2.28. By using exactly the same line of reasoning as in the case of the classical Preisach model (see Section 3 of Chapter I), from these diagrams and (2.86) we derive

$$F(\alpha, \beta, M_1) = \iint\limits_{T(\alpha,\beta)} \mu(\alpha', \beta', M_1)d\alpha' d\beta' = \int_\beta^\alpha \left(\int_{\beta'}^\alpha \mu(\alpha', \beta', M_1)d\alpha' \right) d\beta'. \tag{2.88}$$

By differentiating (2.88) twice, we obtain

$$\mu(\alpha, \beta, M_1) = -\frac{\partial^2 F(\alpha, \beta, M_1)}{\partial\alpha\partial\beta}. \tag{2.89}$$

By recalling the definition (2.87) of $F(\alpha, \beta, M_1)$, from (2.89) we find

$$\mu(\alpha, \beta, M_1) = -\frac{1}{2}\frac{\partial^2 f_{M_1\beta\alpha}}{\partial\alpha\partial\beta}. \tag{2.90}$$

Thus, if the distribution function μ is determined from (2.89) or (2.90), then the restricted Preisach model will match the output increments along second-order reversal curves $f_{M_1\beta\alpha}$ attached to the first-order reversal curve $f_{M_1\beta}$. This implies that the limiting ascending branch f^+ will be matched because this branch coincides with the second-order reversal curves $f_{M_1\beta_0\alpha}$ and because the value f_- is matched by the model. From (2.87) we also find that the output increments along the first-order reversal curve $f_{M_1\beta}$ will be matched because these increments are twice of $F(M_1, \beta, M_1)$. Since the limiting ascending branch, output increments along the first-order transition curves $f_{M_1\beta}$ and the output increment along the second-order reversal curves $f_{M_1\beta\alpha}$ will all be matched, we conclude that the second-order transition curves $f_{M_1\beta\alpha}$ themselves will be matched.

Up to this point it has been tacitly assumed that input variations are started from the state of negative saturation ($u(t) \leq \beta_0$). This is a natural assumption because the state of negative saturation is a well-defined state. However, another well-defined state is the state of positive saturation ($u(t) \geq \alpha_0$). If input variations start from this state, then it is natural to assume that the distribution function depends on the first global minimum m_1. This leads us to the following restricted Preisach model:

$$f(t) = \iint_{T_{m_1}} \tilde{\mu}(\alpha, \beta, m_1)\hat{\gamma}_{\alpha\beta}u(t)d\alpha d\beta + C_{m_1}, \tag{2.91}$$

where the support of $\tilde{\mu}(\alpha, \beta, m_1)$ is the triangle T_{m_1} defined by inequalities $m_1 \leq \beta \leq \alpha \leq \alpha_0$ (see Fig. 2.29).

The model (2.91) can be regarded as a counterpart of the model (2.82) and our immediate goal is to establish some connections between distribution functions μ and $\tilde{\mu}$. To this end, we first find the constant C_{m_1} in (2.91) by matching the output values $f_{m_1}^-$ and f_+:

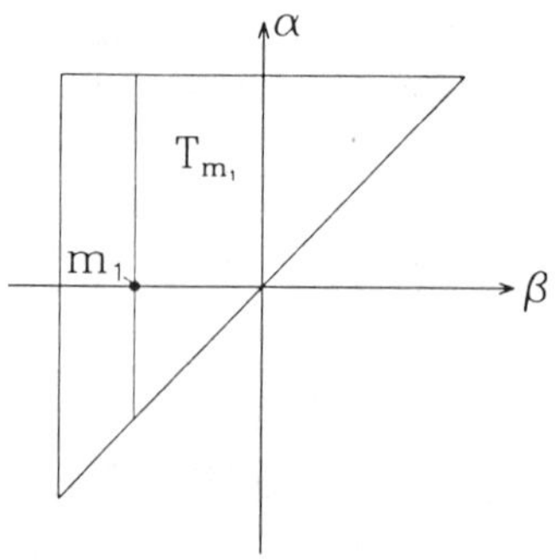

Fig. 2.29

$$f_{m_1}^- = -\iint_{T_{m_1}} \tilde{\mu}(\alpha, \beta, m_1) d\alpha d\beta + C_{m_1}, \tag{2.92}$$

$$f_+ = \iint_{T_{m_1}} \tilde{\mu}(\alpha, \beta, m_1) d\alpha d\beta + C_{m_1}. \tag{2.93}$$

From (2.92) and (2.93), we find

$$C_{m_1} = \frac{f_{m_1}^- + f_+}{2}, \tag{2.94}$$

and the restricted model (2.91) can be represented as

$$f(t) = \iint_{T_{m_1}} \tilde{\mu}(\alpha, \beta, m_1) \hat{\gamma}_{\alpha\beta} u(t) d\alpha d\beta + \frac{f_{m_1}^- + f_+}{2}. \tag{2.95}$$

To determine the distribution function $\tilde{\mu}(\alpha, \beta, m_1)$, we shall use the decreasing second-order reversal curves $f_{m_1\tilde{\alpha}\tilde{\beta}}$ which are attached to the increasing first-order reversal curves $f_{m_1\tilde{\alpha}}$ (see Fig. 2.30). As before we can introduce the function

$$\tilde{F}(\tilde{\alpha}, \tilde{\beta}, m_1) = \tfrac{1}{2}(f_{m_1\tilde{\alpha}} - f_{m_1\tilde{\alpha}\tilde{\beta}}) \tag{2.96}$$

and show that

$$\tilde{F}(\tilde{\alpha}, \tilde{\beta}, m_1) = \iint_{T(\tilde{\alpha}, \tilde{\beta})} \tilde{\mu}(\alpha', \beta') d\alpha' d\beta'. \tag{2.97}$$

From (2.97), we derive

$$\tilde{\mu}(\tilde{\alpha}, \tilde{\beta}, m_1) = -\frac{\partial^2 \tilde{F}(\tilde{\alpha}, \tilde{\beta}, m_1)}{\partial \tilde{\alpha} \partial \tilde{\beta}}. \tag{2.98}$$

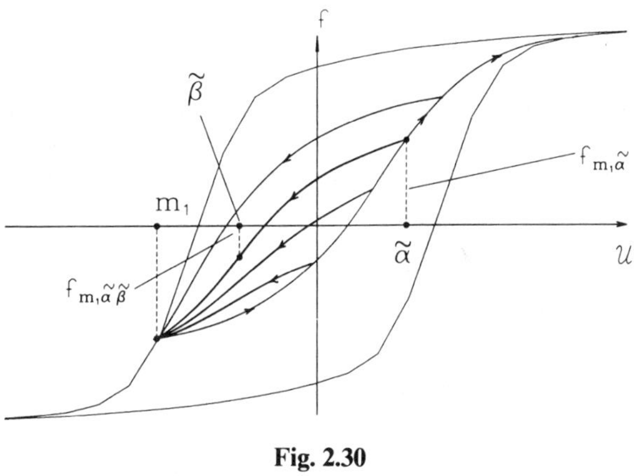

Fig. 2.30

By invoking (2.96) in (2.98), we obtain

$$\tilde{\mu}(\tilde{\alpha}, \tilde{\beta}, m_1) = \frac{1}{2}\frac{\partial^2 f_{m_1\tilde{\alpha}\tilde{\beta}}}{\partial\tilde{\alpha}\partial\tilde{\beta}}. \tag{2.99}$$

Next, we will utilize the symmetry between the ascending and descending second-order reversal curves. In mathematical terms, this symmetry means that if

$$m_1 = -M_1, \qquad \tilde{\alpha} = -\beta, \qquad \tilde{\beta} = -\alpha, \tag{2.100}$$

then

$$\tilde{F}(\tilde{\alpha}, \tilde{\beta}, m_1) = F(\alpha, \beta, M_1). \tag{2.101}$$

The last formula can also be expressed as

$$\tilde{F}(-\beta, -\alpha, -M_1) = F(\alpha, \beta, M_1). \tag{2.102}$$

By substituting (2.101) and (2.100) into (2.98) and taking into account (2.89), we derive

$$\tilde{\mu}(\tilde{\alpha}, \tilde{\beta}, m_1) = -\frac{\partial^2 F(\alpha, \beta, M_1)}{\partial\alpha\partial\beta} = \mu(\alpha, \beta, M_1). \tag{2.103}$$

From (2.100) and (2.103), we finally obtain

$$\mu(\alpha, \beta, M_1) = \tilde{\mu}(-\beta, -\alpha, -M_1). \tag{2.104}$$

The last expression can be regarded as a generalization of the mirror symmetry property (1.46) previously derived for the classical Preisach model. Indeed, if $M_1 = \alpha_0$ and $m_1 = \beta_0$, $(\beta_0 = -\alpha_0)$, then the models (2.86) and (2.95), respectively, coincide with the classical Preisach model. Consequently, we have

$$\mu(\alpha, \beta, \alpha_0) = \mu(\alpha, \beta), \qquad \tilde{\mu}(\alpha, \beta, -\alpha_0) = \mu(\alpha, \beta), \tag{2.105}$$

and the equality (2.104) can be rewritten as

$$\mu(\alpha, \beta) = \mu(-\beta, -\alpha), \tag{2.106}$$

which is the property of mirror symmetry for the classical Preisach model.

The formula (2.104) also shows that, having determined the function $\mu(\alpha, \beta, M_1)$, we have solved the identification problem for the model (2.86) and its counterpart (2.95). This suggests that the model (2.95) is not independent of the model (2.86), but rather complements the latter model. In a way, we deal here with the same model which is written in two different forms. These two different forms correspond to two different initial states (states of positive or negative saturation). For this reason, in our subsequent discussion we will not draw any distinction between the models (2.86) and (2.95). In the sequel, all results will be discussed for the model (2.86), and it will be tacitly assumed that they are also valid for the model (2.95).

It is clear that the model (2.86) has almost identical structure with the classical Preisach model. For this reason, it is apparent that the wiping-out property holds for the restricted model (2.86). However, the congruency property of minor hysteresis loops undergoes some modification. This modification can be described as follows. We call minor hysteresis loops *comparable* if they are formed as a result of back-and-forth input variations between the same consecutive reversal values *and* these input variations take place some time after the same largest input maximum was achieved. The given definition of comparable minor loops implies the possibility of different past input histories between the time when the largest maximum M_1 was achieved and the time when the above-mentoned back-and-forth input variations commence. By using the same reasoning as in Section 2 of Chapter I, we can prove that all comparable minor loops described by the restricted model (2.86) are congruent. It is clear from the above statement that the congruency of minor hysteresis loops prescribed by the model (2.86) for the same input reversal values are to a certain extent history dependent. This is not the case for the classical Preisach model.

It turns out that the wiping-out property and the modified congruency property are characteristic for the restricted model (2.86) in a sense that the following result is valid.

Representation Theorem

The wiping-out property and the congruency property of comparable minor loops constitute necessary and sufficient conditions for the representation of actual hysteresis nonlinearities by the restricted Preisach model (2.86).

The proof of this theorem is very similar to the proof of the representation theorem for the classical Preisach model. For this reason, the proof is omitted.

It is clear that the modified congruency property is more general than the congruency property of the classical Preisach model. Thus, the restricted model (2.86) is more general than the classical one. In addition, the restricted model (2.86) allows one to fit experimentally measured first- and second-order reversal curves, while the classical Preisach model is able to fit only first-order reversal curves. Since higher-order reversal curves are "sandwiched" between first- and second-order reversal curves, it is natural to expect that the restricted model will be more accurate than the classical Preisach model.

We next turn to the discussion of numerical implementation of the restricted model (2.86). As before, we will show that double integration in (2.86) can be completely avoided and that an explicit expression for the output $f(t)$ in terms of the experimentally measured function, $F(\alpha, \beta, M_1)$ can be derived. The derivation proceeds as follows. From (2.86) by means of simple transformations we find

$$\begin{aligned} f(t) &= \iint\limits_{S^+_{M_1}(t)} \mu(\alpha, \beta, M_1) d\alpha d\beta - \iint\limits_{S^-_{M_1}(t)} \mu(\alpha, \beta, M_1) d\alpha d\beta + \frac{f^+_{M_1} + f_-}{2} \\ &= 2 \iint\limits_{S^+_{M_1}(t)} \mu(\alpha, \beta, M_1) d\alpha d\beta - \iint\limits_{T_{M_1}} \mu(\alpha, \beta, M_1) d\alpha d\beta + \frac{f^+_{M_1} + f_-}{2}. \qquad (2.107) \end{aligned}$$

By invoking (2.83) and (2.84), we derive

$$\iint_{T_{M_1}} \mu(\alpha, \beta, M_1)d\alpha d\beta = \frac{f_{M_1}^+ - f_-}{2}. \tag{2.108}$$

By substituting (2.108) into (2.107), we obtain

$$f(t) = 2 \iint_{S_{M_1}^+(t)} \mu(\alpha, \beta, M_1)d\alpha d\beta + f_-. \tag{2.109}$$

The set $S_{M_1}^+(t)$ can be subdivided into n trapezoids Q_k shown in Fig. (2.31). As a result, we have

$$\iint_{S_{M_1}^+(t)} \mu(\alpha, \beta, M_1)d\alpha d\beta = \sum_{k=1}^{n(t)} \iint_{Q_k} \mu(\alpha, \beta M_1)d\alpha d\beta. \tag{2.110}$$

Each integral in the right-hand side of (2.110) can be represented as

$$\iint_{Q_k} \mu(\alpha, \beta, M_1)d\alpha d\beta = \iint_{T(M_k, m_{k-1})} \mu(\alpha, \beta, M_1)d\alpha d\beta - \iint_{T(M_k, m_k)} \mu(\alpha, \beta, M_1)d\alpha d\beta. \tag{2.111}$$

By recalling (2.88), from (2.111) we find

$$\iint_{Q_k} \mu(\alpha, \beta, M_1)d\alpha d\beta = F(M_k, m_{k-1}, M_1) - F(M_k, m_k, M_1). \tag{2.112}$$

By substituting (2.112) into (2.110) and then (2.110) into (2.109), we finally obtain

$$f(t) = 2 \sum_{k=1}^{n(t)} [F(M_k, m_{k-1}, M_1) - F(M_k, m_k, M_1)] + f_-. \tag{2.113}$$

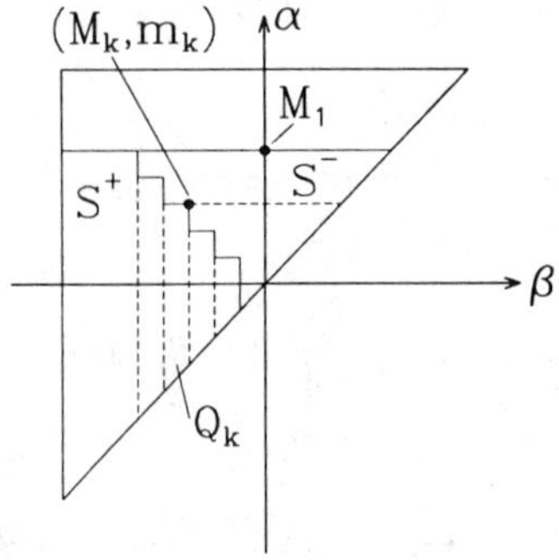

Fig. 2.31

Thus, the formula (2.113) expresses the output $f(t)$ explicitly in terms of experimentally measured second-order reversal curves which are related to the function F by the equation (2.87). This formula has a twofold advantage over the expression (2.86). First, double integration is avoided. Second, the determination of $\mu(\alpha, \beta, M_1)$ through differentiation of experimentally obtained data (see (2.89) or (2.90)) is completely circumvented. This is a welcome feature because the above differentiation may amplify noise inherently present in any experimental data.

So far we have discussed the restricted Preisach model for which the distribution function μ depends only on the first global extrema M_1 or m_1. It is natural to call this a first-order restricted Preisach model. However, further generalizations in this direction are possible. In these generalizations, the function μ is assumed to be dependent on some finite sequence of past dominant extrema: $\mu(\alpha, \beta, M_1, m_1, M_2, m_2, \ldots M_k, m_k)$. It is natural to call these high-order restricted Preisach models. To make the general idea of these models clear, consider first a second-order restricted Preisach model. In this model, the function μ is assumed to be dependent on M_1 and m_1, and the model itself is defined as

$$f(t) = \iint_{T_{M_1 m_1}} \mu(\alpha, \beta, M_1, m_1)\hat{\gamma}_{\alpha\beta} u(t)\, d\alpha\, d\beta + C_{M_1 m_1}, \tag{2.114}$$

where the support of μ is the triangle $T_{M_1 m_1}$ specified by inequalities $m_1 \leq \beta \leq \alpha \leq M_1$ (see Fig. 2.32).

For any fixed extrema M_1 and m_1, the model (2.114) will be used to describe hysteresis behavior in the region confined between the first-order reversal curve $f_{M_1 \lambda}$ and the second-order reversal curve $f_{M_1 m_1 \alpha}$ (see Fig. 2.33). This explains the origin of the term "restricted" used for the name of this model. However, the described restriction does not narrow at all the region of applicability of this model. This is because M_1 and m_1 are free parameters which may assume any values between β_0 and α_0. These values depend on a particular past input history and may even change with time when the previous extrema M_1 and m_1 are wiped out by subsequent input variations.

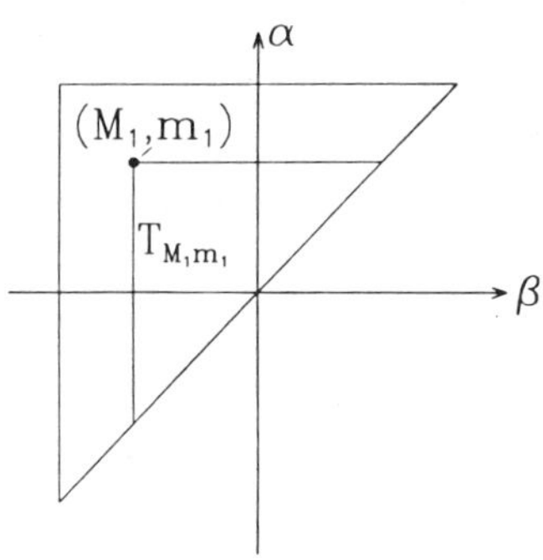

Fig. 2.32

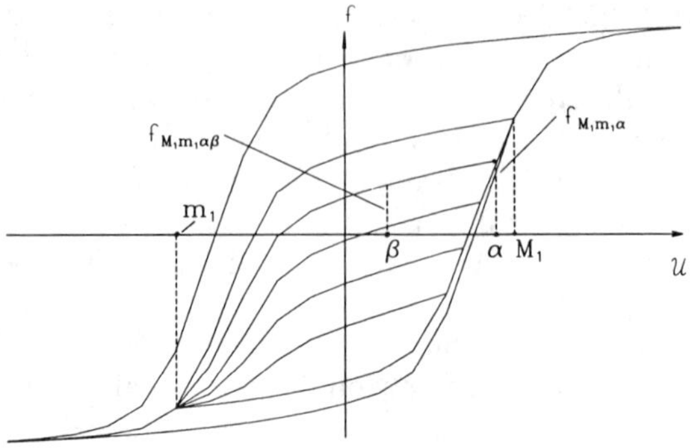

Fig. 2.33

The constant C_{M_1,m_1} in (2.114) can be determined from the condition that the model matches the output values $f_{M_1}^+$ and $f_{M_1m_1}$:

$$f_{M_1}^+ = \iint_{T_{M_1m_1}} \mu(\alpha, \beta, M_1, m_1)d\alpha d\beta + C_{M_1m_1}, \tag{2.115}$$

$$f_{M_1m_1} = -\iint_{T_{M_1m_1}} \mu(\alpha, \beta, M_1, m_1)d\alpha d\beta + C_{M_1m_1}. \tag{2.116}$$

From (2.115) and (2.116), we find

$$C_{M_1m_1} = \frac{f_{M_1}^+ + f_{M_1m_1}}{2}, \tag{2.117}$$

which leads to the following expression for the model (2.114):

$$f(t) = \iint_{T_{M_1m_1}} \mu(\alpha, \beta, M_1, m_1)\hat{\gamma}_{\alpha\beta}u(t)d\alpha d\beta + \frac{f_{M_1}^+ + f_{M_1m_1}}{2}. \tag{2.118}$$

To determine the distribution function $\mu(\alpha, \beta, M_1, m_1)$, the experimentally measured third-order reversal curves $f_{M_1m_1\alpha\beta}$ (see Fig. 2.33) should be employed. By using these curves, we can introduce the function

$$F(\alpha, \beta, M_1, m_1) = \tfrac{1}{2}(f_{M_1m_1\alpha} - f_{M_1m_1\alpha\beta}). \tag{2.119}$$

Exactly in the same way as before it can be shown that the second-order restricted Preisach model will fit the third-order reversal curves if the function $F(\alpha, \beta, M_1, m_1)$ is related to the function $\mu(\alpha, \beta, M_1, m_1)$ by the equation

$$F(\alpha, \beta, M_1, m_1) = \iint_{T(\alpha,\beta)} \mu(\alpha', \beta', M_1, m_1)d\alpha' d\beta'. \tag{2.120}$$

From the last equation we easily derive

$$\mu(\alpha, \beta, M_1, m_1) = -\frac{\partial^2 F(\alpha, \beta, M_1, m_1)}{\partial\alpha\partial\beta} \tag{2.121}$$

and

$$\mu(\alpha, \beta, M_1, m_1) = \frac{1}{2}\frac{\partial^2 f_{M_1 m_1 \alpha\beta}}{\partial\alpha\partial\beta}. \tag{2.122}$$

By its design, the model (2.11) is valid under the condition that the initial state of hysteresis transducer is the state of negative saturation. If the initial state is the state of positive saturation, then the following counterpart of the model (2.111) can be used:

$$f(t) = \iint_{T_{\tilde{M}_1 \tilde{m}_1}} \tilde{\mu}(\alpha, \beta, \tilde{m}_1, \tilde{M}_1)\hat{\gamma}_{\alpha\beta}u(t)\,d\alpha\,d\beta + \frac{f_{\tilde{m}_1}^{-} + f_{\tilde{m}_1 \tilde{M}_1}}{2}, \tag{2.123}$$

where $\tilde{m}_1$ is a global minimum and $\tilde{M}_1$ is a subsequent global maximum of the input.

As before, we can establish the following relationship between the function μ and $\tilde{\mu}$:

$$\mu(\alpha, \beta, M_1, m_1) = \tilde{\mu}(-\beta, -\alpha, -M_1, -m_1), \tag{2.124}$$

which shows that (2.118) and (2.123) are just different forms of the same model. These two different forms correspond to two different well-defined initial states of hysteresis transducer.

It is apparent that the congruency property of minor hysteresis loops undergoes further modification (further relaxation) in the case of the second-order restricted Preisach model (2.118). The essence of this modification is that for minor loops to be congruent the corresponding inputs must vary back-and-forth between the same consecutive reversal values and also assume in the past the same values of M_1 and m_1. It can be proven that the wiping-out property and the above-mentioned modified congruency property of minor loops constitute necessary and sufficient conditions for the representation of actual hysteresis nonlinearities by the model (2.118). This is the so-called representation theorem for the second-order restricted Preisach model. As before, it can be shown that the following expression can be used for the numerical implementation of the model (2.118):

$$f(t) = 2\sum_{k}^{n(t)} [F(M_k, m_{k-1}, M_1, m_1) - F(M_k, m_k, M_1, m_1)] + f_{M_1 m_1}. \tag{2.125}$$

The main advantage of this expression over the formula (2.118) is that it represents the output $f(t)$ directly in terms of the experimentally measured third-order transition curves.

It is clear from the previous discussion that higher-order restricted Preisach models of hysteresis can be defined as

$$f(t) = \iint_{T_{M_1 m_1 \cdots M_k}} \mu(\alpha, \beta, M_1, m_1, \ldots M_k)\hat{\gamma}_{\alpha\beta}u(t)d\alpha d\beta + C_{M_1 m_1 \ldots M_k}. \quad (2.126)$$

It is also apparent now how higher-order transition curves can be used for the determination of the function $\mu(\alpha, \beta, M_1, m_1, \ldots M_k)$. It goes without saying that by increasing the order of the restricted Preisach model we can increase the accuracy of this model. However, this increase in accuracy is amply paid for by the increase in the amount of experimental data required for the identification of higher-order models. For this reasons, the use of restricted Preisach models of order higher than two does not seem to be practically feasible or attractive.

We conclude this section by a brief discussion of the restricted nonlinear Preisach model of hysteresis.

It is clear from the name of this model that it combines the main features of restricted and nonlinear models. The first-order restricted nonlinear (input-dependent) Preisach model can be defined as

$$f(t) = \iint_{R_{M_1 u}} \mu(\alpha, \beta, u(t), M_1)\hat{\gamma}_{\alpha\beta}u(t)d\alpha d\beta + \frac{f_{M_1 u(t)} + f^+_{u(t)}}{2}, \quad (2.127)$$

where the moving support of $\mu(\alpha, \beta, u(t), M_1)$ is the rectangle $R_{M_1 u}$ specified by inequalities $\beta_0 \leq \beta \leq u(t)$, $u(t) \leq \alpha \leq M_1$ (see Fig. 2.34), and the meaning of $f_{M_1 u(t)}$ is clear from the Fig. 2.35.

By introducing the function

$$P(\alpha, \beta, u, M_1) = f_{M_1 \beta\alpha u} - f_{M_1 \beta u} \quad (2.128)$$

and by using the same reasoning as in the previous section, we can establish that

$$P(\alpha, \beta, u, M_1) = 2 \iint_{R(\alpha, \beta, u)} \mu(\alpha', \beta', u, M_1)d\alpha' d\beta'. \quad (2.129)$$

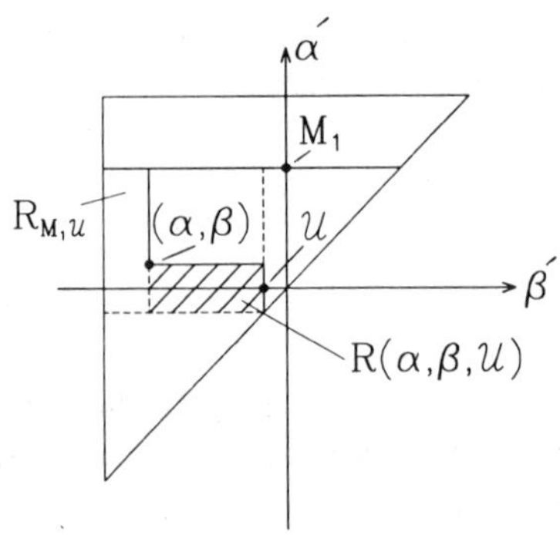

Fig. 2.34

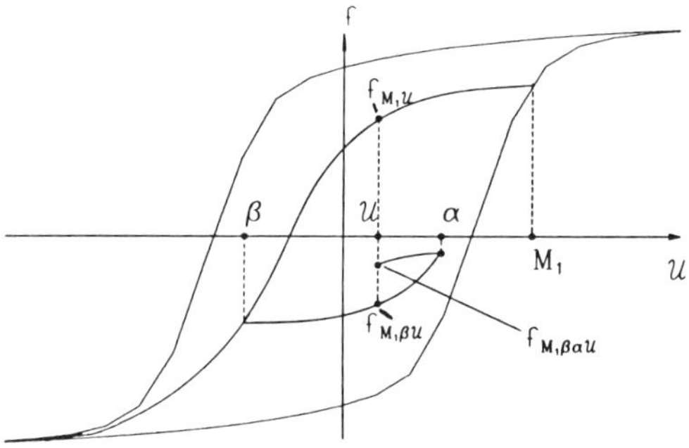

Fig. 2.35

From (2.129) we can derive the following expressions:

$$\mu(\alpha, \beta, u, M_1) = -\frac{1}{2}\frac{\partial^2 P(\alpha, \beta, u, M_1)}{\partial\alpha\partial\beta} \tag{2.130}$$

and

$$\mu(\alpha, \beta, u, M_1) = -\frac{1}{2}\frac{\partial^2 f_{M_1\beta\alpha u}}{\partial\alpha\partial\beta}. \tag{2.131}$$

We can also derive the following explicit expression for the output $f(t)$ of the model (2.127):

$$f(t) = f_{M_1u(t)} + \sum_{k}^{n(t)} [P(M_{k+1}, m_k, u(t), M_1) - P(M_k, m_k, u(t), M_1)]. \tag{2.132}$$

It is worthwhile to remember here that the function $P(\alpha, \beta, u, M_1)$ is directly related to experimental data and it has the meaning of output increments between the second- and third-order reversal curves.

It is easy to formulate and to prove the representation theorem for the model (2.127). This is left to the reader as a useful exercise.

4 Dynamic Preisach Models of Hysteresis

In all previous sections our discussion has been centered around the classical and generalized Preisach models of hysteresis which are static in nature. The term "static" implies that in these models only past input extrema leave their mark upon the future values of output, while the speed of input and output variations has no influence on branching. The intent of this section is to relax the "static property" of Preisach-type models. For this reason, new Preisach-type models which are applicable to the description of dynamic hysteresis are introduced. The identification problem of fitting these models to some experimental data is then studied. Finally, some discussion is presented concerning numerical implementation of the dynamic Preisach models of hysteresis.

The main idea behind the dynamic Preisach-type models of hysteresis is to introduce the dependence of μ-functions on the speed of output variations, df/dt. This leads to the following dynamic Preisach models of hysteresis:

$$f(t) = \iint_{\alpha \geq \beta} \mu\left(\alpha, \beta, \frac{df}{dt}\right) \hat{\gamma}_{\alpha\beta} u(t) d\alpha d\beta, \tag{2.133}$$

$$f(t) = \iint_{R_{u(t)}} \mu\left(\alpha, \beta, u(t), \frac{df}{dt}\right) \hat{\gamma}_{\alpha\beta} u(t) d\alpha d\beta + \frac{f^-_{u(t)} + f^+_{u(t)}}{2}. \tag{2.134}$$

The above models are "dynamic" generalizations of the classical Preisach model and the nonlinear Preisach model, respectively. Similar generalizations for the restricted Preisach models are apparent. They will not be discussed in this section because their treatment strongly parallels that of the model (2.133).

The direct utilization of the models (2.133) and (2.134) is associated with some untractable difficulties. First, μ-functions depend on the unknown quantity df/dt, and this complicates numerical implementations of the models (2.133) and (2.134). Second, it is not clear how to pose the identification problems for these models. The above difficulties can be completely circumvented by using the power series expansions for μ-functions with respect to df/dt:

$$\mu\left(\alpha, \beta, \frac{df}{dt}\right) = \mu_0(\alpha, \beta) + \frac{df}{dt}\mu_1(\alpha, \beta) + \cdots, \tag{2.135}$$

$$\mu\left(\alpha, \beta, u(t), \frac{df}{dt}\right) = \mu_0(\alpha, \beta, u(t)) + \frac{df}{dt}\mu_1(\alpha, \beta, u(t)) + \cdots. \tag{2.136}$$

By retaining only the two first terms of the above expansion, we arrive at the following dynamic models:

$$f(t) = \iint_{\alpha \geq \beta} \mu_0(\alpha, \beta) \hat{\gamma}_{\alpha\beta} u(t) d\alpha d\beta + \frac{df}{dt} \iint_{\alpha \geq \beta} \mu_1(\alpha, \beta) \hat{\gamma}_{\alpha\beta} u(t) d\alpha d\beta, \tag{2.137}$$

$$f(t) = \iint_{R_{u(t)}} \mu_0(\alpha, \beta, u(t)) \hat{\gamma}_{\alpha\beta} u(t) d\alpha d\beta + \frac{f^-_{u(t)} + f^+_{u(t)}}{2}$$

$$+ \frac{df}{dt} \iint_{R_{u(t)}} \mu_1(\alpha, \beta u(t)) \hat{\gamma}_{\alpha\beta} u(t) d\alpha d\beta. \tag{2.138}$$

It is clear that in the case of very slow output variations, the above models should be reduced to the corresponding static hysteresis models. This means the μ_0-functions in (2.137) and (2.138) should coincide with the μ-functions of static models (2.1) and (2.26), respectively. In other words, the μ_0-functions in (2.137) and (2.138) can be determined by matching static first-order and second-order transition curves, respectively.

The above reasoning suggests that the models (2.137) and (2.138) can be represented in the following equivalent forms:

$$f(t) = \tilde{f}(t) + \frac{df}{dt} \iint_{\alpha \geq \beta} \mu_1(\alpha, \beta) \hat{\gamma}_{\alpha\beta} u(t) d\alpha d\beta, \tag{2.139}$$

$$f(t) = \tilde{f}(t) + \frac{df}{dt} \iint_{R_{u(t)}} \mu_1(\alpha, \beta, u(t)) \hat{\gamma}_{\alpha\beta} u(t) d\alpha d\beta, \tag{2.140}$$

where the $\tilde{f}$-terms stand for the static components of hysteresis nonlinearities:

$$\tilde{f}(t) = \iint_{\alpha \geq \beta} \mu_0(\alpha, \beta) \hat{\gamma}_{\alpha\beta} u(t) d\alpha d\beta, \tag{2.141}$$

$$\tilde{f}(t) = \iint_{R_{u(t)}} \mu_0(\alpha, \beta, u(t)) \hat{\gamma}_{\alpha\beta} u(t) d\alpha d\beta + + \frac{f^+_{u(t)} + f^-_{u(t)}}{2}. \tag{2.142}$$

The expressions (2.139) and (2.140) are transparent from the physical point of view. They show that the instant speeds of output variations are directly proportional to the differences between instant and static output values.

We next turn to the identification problems of determining the μ_1-functions by fitting the models (2.139) and (2.140) to some experimental data. We first consider the identification problem for the model (2.139). The following experiments are used to solve this problem. Starting from the state of negative saturation, the input $u(t)$ is monotonically increased until it reaches some value α at $t = t_0$ and it is kept constant for $t \geq t_0$. As the input is being kept constant, the output relaxes from its value f_α at $t = t_0$ to its static value $\tilde{f}_\alpha$. According to the model (2.139), this relaxation process is described by the differential equation

$$\tau_\alpha \frac{df}{dt} + f = \tilde{f}_\alpha, \tag{2.143}$$

where

$$\tau_\alpha = \iint_{S^-_\alpha} \mu_1(\alpha', \beta') d\alpha' d\beta' - \iint_{S^+_\alpha} \mu_1(\alpha', \beta') d\alpha' d\beta', \tag{2.144}$$

and S^-_α and S^+_α are negative and positive sets on the Preisach diagram (see Fig. 2.36) corresponding to the above input variation.

The solution to the equation (2.143) is given by

$$f(t) = (f_\alpha - \tilde{f}_\alpha) e^{-t/\tau_\alpha} + \tilde{f}_\alpha. \tag{2.145}$$

Thus, τ_α has the meaning of relaxation time and can be experimentally measured.

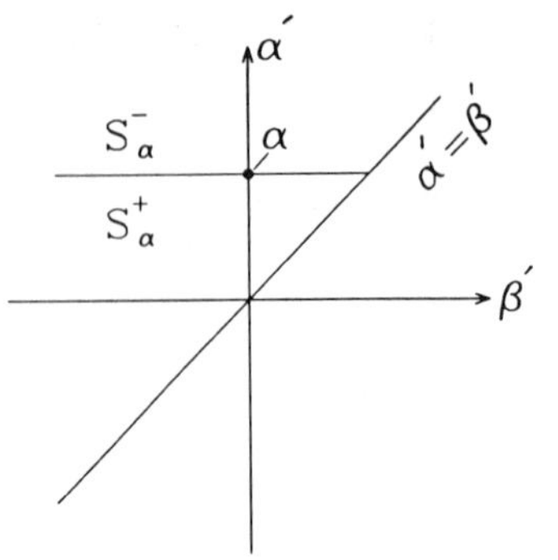

Fig. 2.36

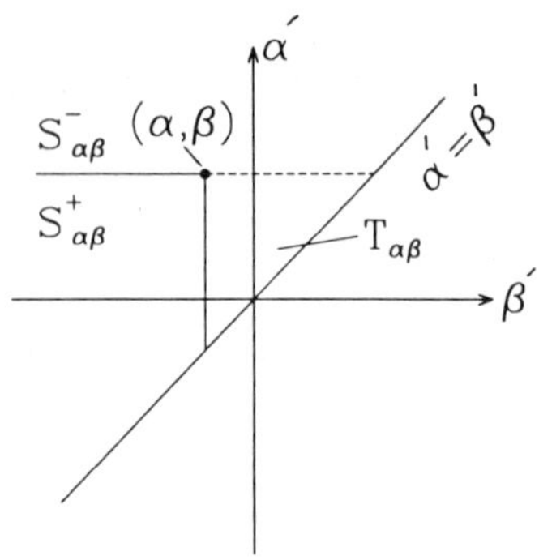

Fig. 2.37

Next, the hysteresis nonlinearity is brought back to the state of negative saturation. Starting from this state, the input is again monotonically increased until it reaches the value α. Then the input is monotonically decreased until it reaches some value β at time $t = t_0'$ and it is kept constant for $t > t_0'$. As the input is being kept constant, the output relaxes from its values $f_{\alpha\beta}$ at $t = t_0'$ to its static value $\tilde{f}_{\alpha\beta}$. The model (2.139) yields the following differential equation for the above relaxation process:

$$\tau_{\alpha\beta} \frac{df}{dt} + f = \tilde{f}_{\alpha\beta}, \tag{2.146}$$

where according to the model (2.139) and the Preisach diagram shown in Fig. 2.37, the coefficient $\tau_{\alpha\beta}$ is given by the expression

$$\tau_{\alpha\beta} = \iint\limits_{S_{\alpha\beta}^-} \mu_1(\alpha', \beta')d\alpha' d\beta' - \iint\limits_{S_{\alpha\beta}^+} \mu_1(\alpha', \beta')d\alpha' d\beta'. \tag{2.147}$$

By solving (2.146), we find

$$f(t) = (f_{\alpha\beta} - \tilde{f}_{\alpha\beta})e^{-t/\tau_{\alpha\beta}} + \tilde{f}_{\alpha\beta}. \tag{2.148}$$

Thus, $\tau_{\alpha\beta}$ has also the meaning of relaxation time and can be experimentally measured. It is apparent that the relaxation time τ_α can be construed as the relaxation time $\tau_{\alpha\alpha}$, and, consequently, it belongs to the set $\{\tau_{\alpha\beta}\}$. We next show that by knowing these relaxation times, we can find the function $\mu_1(\alpha, \beta)$. To this end, we introduce the function

$$q(\alpha, \beta) = \tau_\alpha - \tau_{\alpha\beta}. \tag{2.149}$$

From (2.144), (2.147) and (2.149), we find

$$q(\alpha, \beta) = 2 \iint\limits_{T_{\alpha\beta}} \mu_1(\alpha', \beta')d\alpha' d\beta', \tag{2.150}$$

where $T_{\alpha\beta}$ is a triangle shown in Fig. 2.37.

From (2.150), we can easily derive the expression

$$\mu_1(\alpha, \beta) = -\frac{1}{2}\frac{\partial^2 q_{\alpha\beta}}{\partial\alpha\partial\beta}. \tag{2.151}$$

By recalling (2.149), from (2.151), we find

$$\mu_1(\alpha, \beta) = \frac{1}{2}\frac{\partial^2 \tau_{\alpha\beta}}{\partial\alpha\partial\beta}. \tag{2.152}$$

The solution to the identification problem for the input-dependent (nonlinear) model (2.140) can be found in a similar way. For this reason, our comments will be concise. We consider two types of relaxation processes. A first type process occurs after monotonic input increase to some value α and subsequent monotonic decrease of input to some value u. According to the model (2.140), the first type relaxation processes are described by the differential equation

$$\tau_{\alpha u}\frac{df}{dt} + f = \tilde{f}_{\alpha u},$$

where

$$\tau_{\alpha u} = \iint\limits_{S^-_{\alpha u}} \mu_1(\alpha', \beta', u)\,d\alpha' d\beta' - \iint\limits_{S^+_{\alpha u}} \mu_1(\alpha', \beta', u)\,d\alpha' d\beta'. \tag{2.153}$$

From the previous differential equation, we find

$$f(t) = (f_{\alpha u} - \tilde{f}_{\alpha u})e^{-t/\tau_{\alpha u}} + \tilde{f}_{\alpha u}. \tag{2.154}$$

A second type relaxation process occurs when the input is first monotonically increased to α, then decreased to β, again monotonically increased to u, and kept constant afterwards. These relaxation processes are described by the equation

$$\tau_{\alpha\beta u}\frac{df}{dt} + f = \tilde{f}_{\alpha\beta u}, \tag{2.155}$$

where

$$\tau_{\alpha\beta u} = \iint\limits_{S^-_{\alpha\beta u}} \mu_1(\alpha', \beta', u)\,d\alpha' d\beta' - \iint\limits_{S^+_{\alpha\beta u}} \mu_1(\alpha', \beta', u)\,d\alpha' d\beta'. \tag{2.156}$$

As before, we have

$$f(t) = (f_{\alpha\beta u} - \tilde{f}_{\alpha\beta u})e^{-t/\tau_{\alpha\beta u}} + \tilde{f}_{\alpha\beta u}. \tag{2.157}$$

It is clear from (2.154) and (2.157) that $\tau_{\alpha u}$ and $\tau_{\alpha\beta u}$ have the physical meaning of relaxation times and can be measured experimentally. Knowing these relaxation times, we can define the function

$$Q(\alpha, \beta, u) = \tau_{\alpha u} - \tau_{\alpha\beta u}. \tag{2.158}$$

From (2.153), (2.156) and (2.158), we derive

$$Q(\alpha, \beta, u) = 2 \iint\limits_{R(\alpha,\beta,u)} \mu_1(\alpha', \beta', u)\, d\alpha' d\beta'. \tag{2.159}$$

From (2.159), we obtain

$$\mu_1(\alpha, \beta, u) = -\frac{1}{2} \frac{\partial^2 Q(\alpha, \beta, u)}{\partial\alpha\partial\beta}. \tag{2.160}$$

By invoking (2.158), from (2.160) we have the alternative expression for μ_1:

$$\mu_1(\alpha, \beta, u) = \frac{1}{2} \frac{\partial^2 \tau_{\alpha\beta u}}{\partial\alpha\partial\beta}. \tag{2.161}$$

Consider the relaxation processes which are symmetric to those discussed above. Suppose that we start from the state of positive saturation and reduce the input to some value $\tilde{\beta}$, then increase the input to some value $\tilde{\alpha}$, and keep it constant thereafter. Suppose also that the subsequent output variations are characterized by the relaxation times $\tau_{\tilde{\beta}\tilde{\alpha}}$. If

$$\tilde{\beta} = -\alpha, \qquad \tilde{\alpha} = -\beta, \tag{2.162}$$

then due to the symmetry we have

$$\tau_{\tilde{\beta}\tilde{\alpha}} = -\tau_{\alpha\beta}. \tag{2.163}$$

As before, we can derive that

$$\mu_1(\tilde{\alpha}, \tilde{\beta}) = -\frac{1}{2} \frac{\partial^2 \tau_{\tilde{\beta}\tilde{\alpha}}}{\partial\tilde{\alpha}\partial\tilde{\beta}}. \tag{2.164}$$

By substituting (2.163) and (2.162) into the right-hand side of (2.164) and by recalling (2.152), we obtain

$$\mu_1(\tilde{\alpha}, \tilde{\beta}) = -\frac{1}{2} \frac{\partial^2 \tau_{\alpha\beta}}{\partial\beta\partial\alpha} = \mu_1(\alpha, \beta). \tag{2.165}$$

From (2.165) and (2.162), we conclude

$$\mu_1(-\beta, -\alpha) = \mu_1(\alpha, \beta). \tag{2.166}$$

Similarly, we can introduce the relaxation $\tau_{\tilde{\beta}\tilde{\alpha}\tilde{u}}$ and prove that

$$\mu_1(\tilde{\alpha}, \tilde{\beta}, \tilde{u}) = -\frac{1}{2} \frac{\partial^2 \tau_{\tilde{\beta}\tilde{\alpha}\tilde{u}}}{\partial\tilde{\alpha}\partial\tilde{\beta}}. \tag{2.167}$$

If

$$\tilde{\beta} = -\alpha, \qquad \tilde{\alpha} = -\beta, \qquad \tilde{u} = -u, \tag{2.168}$$

then due to the symmetry we have

$$\tau_{\tilde{\beta}\tilde{\alpha}\tilde{u}} = -\tau_{\alpha\beta u}. \tag{2.169}$$

By using the same line of reasoning as before, from (2.161), (2.167), (2.168), and (2.169) we derive

$$\mu_1(-\beta, -\alpha, -u) = \mu_1(\alpha, \beta, u). \tag{2.170}$$

We next proceed to the numerical implementation of the dynamic models. The models (2.139) and (2.140) can be represented as the following differential equation:

$$\hat{a}(u(t))\frac{df}{dt} + f(t) = \tilde{f}(t), \tag{2.171}$$

where the hysteretic coefficient $\hat{a}(t)$ is defined by the formulas

$$\hat{a}(u(t)) = -\iint_{\alpha \geq \beta} \mu_1(\alpha, \beta)\hat{\gamma}_{\alpha\beta}u(t)d\alpha d\beta, \tag{2.172}$$

$$\hat{a}(u(t)) = -\iint_{R_{u(t)}} \mu_1(\alpha, \beta, u(t))\hat{\gamma}_{\alpha\beta}u(t)d\alpha d\beta \tag{2.173}$$

for the models (2.139) and (2.173), respectively.

The explicit solution to the equation (2.171) is well known and can be expressed as

$$f(t) = b(t)\left[f_0 + \int_0^t \frac{\tilde{f}(\xi)}{a(\xi)b(\xi)}d\xi\right], \tag{2.174}$$

where f_0 is the initial output value and $b(t)$ is given by the formula

$$b(t) = \exp\left(-\int_0^t \frac{d\xi}{a(\xi)}\right). \tag{2.175}$$

The expressions (2.174) and (2.175) can be efficiently used for computing $f(t)$, if $\tilde{f}(t)$ and $a(t)$ are known. The calculation of static $\tilde{f}$-components in the case of the classical Preisach model as well as the nonlinear Preisach model has been discussed in detail in the previous sections. Since the mathematical structure of $a(t)$ is similar to the mathematical structure of the static Preisach models, similar explicit formulas can be derived for the numerical evaluation of $a(t)$. In particular, it can be shown that the following formulas are valid for $a(t)$ in the case of models (2.139) and (2.140), respectively:

$$a(u(t)) = -\frac{1}{2}q(\alpha_0, \beta_0) + \sum_{k=1}^{n(t)} [q(M_k, m_{k-1}) - q(M_k, m_k)], \tag{2.176}$$

$$\hat{a}(u(t)) = \frac{1}{2}Q(\alpha_0, \beta_0, u(t)) + \sum_{k=1}^{n(t)} [Q(M_{k+1}, m_k, u(t)) - Q(M_k, m_k, u(t))]. \tag{2.177}$$

The above formulas are convenient not only because they give explicit expressions for integrals in (2.172) and (2.173), but also because these expressions are presented in terms of experimentally measured data.

It has been tacitly assumed in the foregoing discussion that the relaxation processes used in the identification procedures are well characterized by single relaxation times. If this is not the case and several relaxation times have to be employed to describe the above relaxation processes, then the discussed dynamic models must be generalized. The natural way to generalize these models is to use higher-order differential equations with hysteretic coefficient in order to account for several relaxation times. We demonstrate such a generalization for the second-order dynamic model:

$$\hat{a}^{(2)}(u(t))\frac{d^2f}{dt^2} + \hat{a}^{(1)}(u(t))\frac{df}{dt} + f(t) = \tilde{f}(t), \tag{2.178}$$

where

$$\tilde{f}(t) = \iint_{\alpha \geq \beta} \mu_0(\alpha, \beta)\hat{\gamma}_{\alpha\beta}u(t)d\alpha d\beta, \tag{2.179}$$

$$\hat{a}^{(1)}(u(t)) = \iint_{\alpha \geq \beta} \mu_1(\alpha, \beta)\hat{\gamma}_{\alpha\beta}u(t)d\alpha d\beta, \tag{2.180}$$

$$\hat{a}^{(2)}(u(t)) = \iint_{\alpha \geq \beta} \mu_2(\alpha, \beta)\hat{\gamma}_{\alpha\beta}u(t)d\alpha d\beta. \tag{2.181}$$

To find $\mu_1(\alpha, \beta)$ and $\mu_2(\alpha, \beta)$, we shall use the same two relaxation processes as for the identification of the model (2.139). In the first process, we start from the state of negative saturation and monotonically increase input to some value α and keep it constant thereafter. As the input is being kept constant, the output relaxes. According to (2.178), this relaxation is described by the differential equation

$$a_\alpha^{(2)}\frac{d^2f}{dt^2} + a_\alpha^{(1)}\frac{df}{dt} + f = \tilde{f}_\alpha. \tag{2.182}$$

A solution to this equation has the form

$$f(t) = C_\alpha^{(1)}e^{-t/\tau_\alpha^{(1)}} + C_\alpha^{(2)}e^{-t/\tau_\alpha^{(2)}} + \tilde{f}_\alpha, \tag{2.183}$$

where $\tau_\alpha^{(1)}$ and $\tau_\alpha^{(2)}$ are the roots of the characteristic equation

$$a_\alpha^{(2)}\tau^2 + a_\alpha^{(1)}\tau + 1 = 0. \tag{2.184}$$

Consequently,

$$a_\alpha^{(2)} = \frac{1}{\tau_\alpha^{(1)}\tau_\alpha^{(2)}}, \tag{2.185}$$

$$a_\alpha^{(1)} = -\frac{\tau_\alpha^{(1)} + \tau_\alpha^{(2)}}{\tau_\alpha^{(1)}\tau_\alpha^{(2)}}. \tag{2.186}$$

The constants $\tau_\alpha^{(1)}$ and $\tau_\alpha^{(2)}$ have the physical meaning of relaxation times and can be measured experimentally. This leads to the experimental determination of $a_\alpha^{(2)}$ and $a_\alpha^{(1)}$ according to (2.185) and (2.186).

Now we consider the second relaxation process. In this case we start from the state of negative saturation, increase input to some value α, then decrease input to some value β, and keep it constant thereafter. As the input is being kept constant, the output relaxes and this relaxation is governed by the equation

$$a_{\alpha\beta}^{(2)}\frac{d^2f}{dt^2} + a_{\alpha\beta}^{(1)}\frac{df}{dt} + f = \tilde{f}_{\alpha\beta}, \tag{2.187}$$

whose solution is given by

$$f(t) = C_{\alpha\beta}^{(1)}e^{-t/\tau_{\alpha\beta}^{(1)}} + C_{\alpha\beta}^{(2)}e^{-t/\tau_{\alpha\beta}^{(2)}} + \tilde{f}_{\alpha\beta}. \tag{2.188}$$

The constants $\tau_{\alpha\beta}^{(1)}$ and $\tau_{\alpha\beta}^{(2)}$ have the meaning of relaxation times and can be measured experimentally. As soon as this is done, we can find $a_{\alpha\beta}^{(2)}$ and $a_{\alpha\beta}^{(1)}$:

$$a_{\alpha\beta}^{(2)} = \frac{1}{\tau_{\alpha\beta}^{(1)}\tau_{\alpha\beta}^{(2)}}, \tag{2.189}$$

$$a_{\alpha\beta}^{(1)} = -\frac{\tau_{\alpha\beta}^{(1)} + \tau_{\alpha\beta}^{(2)}}{\tau_{\alpha\beta}^{(1)}\tau_{\alpha\beta}^{(2)}}. \tag{2.190}$$

Next we introduce the functions

$$q_{\alpha\beta}^{(1)} = a_\alpha^{(1)} - a_{\alpha\beta}^{(1)}, \tag{2.191}$$

$$q_{\alpha\beta}^{(2)} = a_\alpha^{(2)} - a_{\alpha\beta}^{(1)}, \tag{2.192}$$

which are directly related to the above-mentioned experimental data. Using the same reasoning which we have used many times before, we can establish the formulas

$$q_{\alpha\beta}^{(1)} = 2\iint\limits_{T(\alpha,\beta)} \mu_1(\alpha', \beta')\,d\alpha'\,d\beta', \tag{2.193}$$

$$q_{\alpha\beta}^{(2)} = 2\iint\limits_{T(\alpha,\beta)} \mu_2(\alpha', \beta')\,d\alpha'\,d\beta', \tag{2.194}$$

from which we derive

$$\mu_1(\alpha, \beta) = -\frac{1}{2}\frac{\partial^2 q_{\alpha\beta}^{(1)}}{\partial\alpha\,\partial\beta}, \tag{2.195}$$

$$\mu_2(\alpha, \beta) = -\frac{1}{2}\frac{\partial^2 q_{\alpha\beta}^{(2)}}{\partial\alpha\,\partial\beta}. \tag{2.196}$$

Thus, the identification problem for the second-order model (2.178)–(2.181) is solved. Extensions to higher-order dynamic models are straightforward.

5 Preisach Model of Hysteresis with Accommodation

The Preisach-type models described in previous sections exhibit the wiping-out property. This property results in an immediate formation of minor hysteresis loops after only one cycle of back-and-forth input variations between any two consecutive extremum values. However, experiments show that hysteresis loop formations are often preceded by some stabilization process which may require an appreciable number of cycles before a stable minor loop is achieved. This stabilization process is often called in the literature the "accommodation" or "reptation" process. Sometimes this accommodation process can be appreciable and then it is important to model it. In this section we discuss a certain modification of the moving Preisach model which allows one to account for the accommodation process.

We define the Preisach model with accommodation as

$$f(t) = \iint_{R_{u(t)}} \mu(\alpha, \beta, f^{(m)})\hat{\gamma}_{\alpha\beta}u(t)\,d\alpha\, d\beta + r(u(t)). \tag{2.197}$$

In the previous formula $r(u(t))$ stands for a fully reversible component which is represented by some single-valued function of $u(t)$, while

$$f^{(m)} = M_k^f \quad \text{if} \quad (\alpha, \beta) \in R_k(t), \tag{2.198}$$

where M_k^f are local extremum values of output f and $R_k(t)$ are rectangular regions formed after the extremum M_k^f had been achieved (see Fig. 2.38).

It is easy to see that the regions $R_k(t)$ are reduced with time and can even be completely wiped out. This fact is reflected in (2.198) by indicating that R_k are dependent of t. It is also clear from Fig. 2.38 that M_k^f are local output minima for even k and local maxima for odd k.

We next prove that the model (2.197) does account for the accommodation process. For the sake of clarity, we will show this for the simplest case when the initial state of a hysteresis transducer is the state of negative saturation,

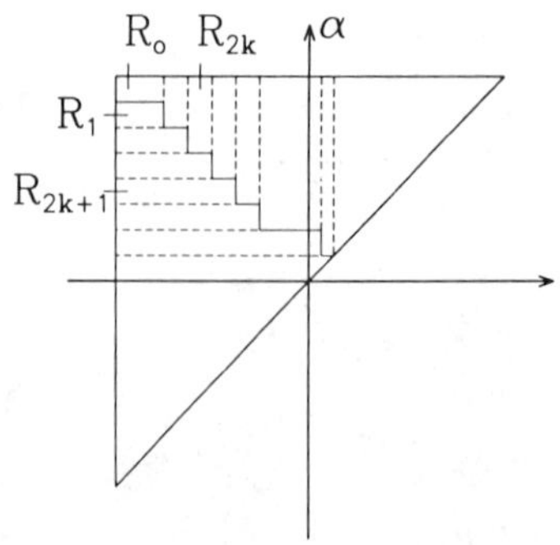

Fig. 2.38

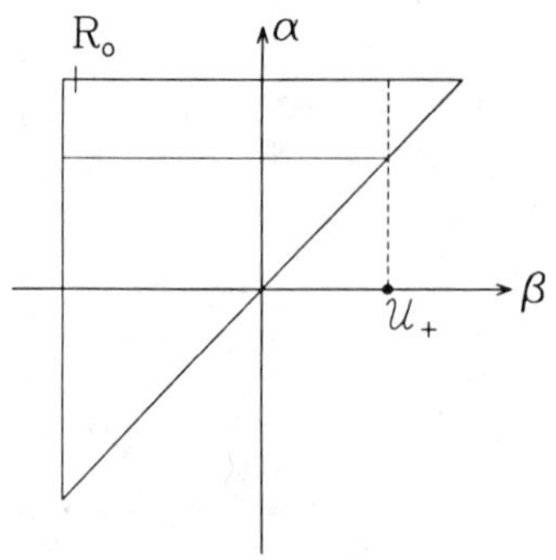

Fig. 2.39

and the input $u(t)$ is monotonically increased to some value u_+ and is being varied back-and-forth between u_+ and u_- thereafter. More general cases can be treated in a similar way.

As the input is increased to the value u_+, the output reaches maximum value M_1^f which, according to (2.197) and Fig. 2.39, is equal to

$$M_1^f = -\iint_{R_0} \mu(\alpha, \beta, M_0^f) d\alpha d\beta + r(u_+), \tag{2.199}$$

where

$$M_0^f = f_-, \tag{2.200}$$

and f_- is the output value in the state of negative saturation.

As the input is monotonically reduced to the value u_-, the output reaches minimum value M_2^f. According to (2.197) and Fig. 2.40, this minimum value is given by

$$M_2^f = \iint_{R_1} \mu(\alpha, \beta, M_1^f) d\alpha d\beta - \iint_{R_0} \mu(\alpha, \beta, M_0^f) d\alpha d\beta + r(u_-). \tag{2.201}$$

It is clear that

$$M_2^f \neq M_0^f = f_-. \tag{2.202}$$

As the input is monotonically increased again to the value u_+, the output reaches some maximum value M_3^f. According to (2.197) and Fig. 2.41, for this value we find

$$M_3^f = -\iint_{R_0} \mu(\alpha, \beta, M_0^f) d\alpha d\beta - \iint_{R_2} \mu(\alpha, \beta, M_2^f) d\alpha d\beta + r(u_+). \tag{2.203}$$

By comparing Figs. 2.39 and 2.41 as well as formulas (2.199) and (2.203) and by taking the inequality (2.202) into account, we conclude

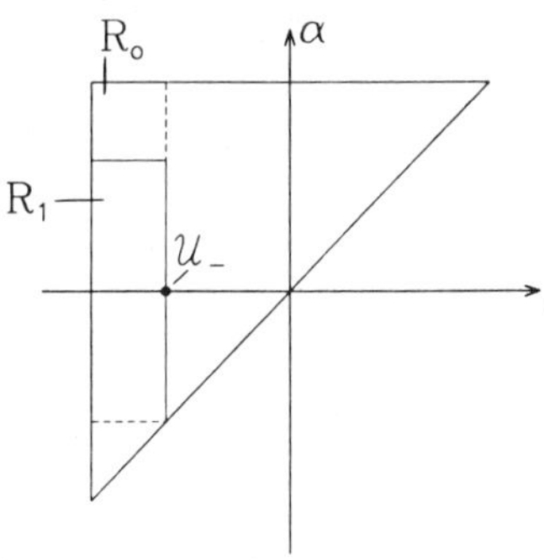

Fig. 2.40

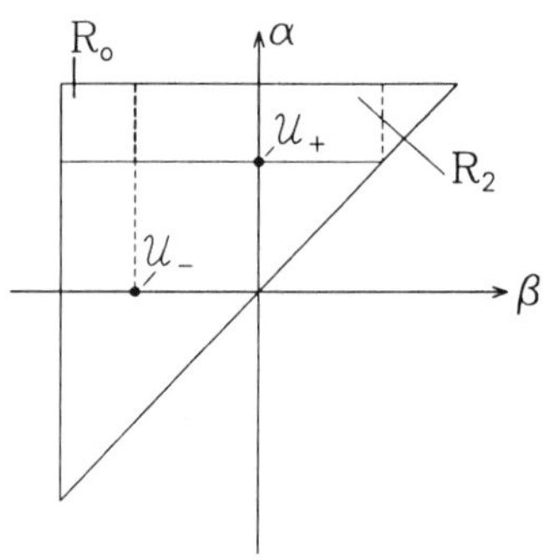

Fig. 2.41

$$M_3^f \neq M_1^f. \tag{2.204}$$

As the input is monotonically decreased again to the value u_-, the output reaches some minimum value M_4^f which according to Fig. 2.42 is given by

$$M_4^f = \iint_{R_3} \mu(\alpha, \beta, M_3^f)\,d\alpha\,d\beta - \iint_{R_0} \mu(\alpha, \beta, M_0^f)\,d\alpha\,d\beta + r(u_-). \tag{2.205}$$

By comparing formulas (2.201) and (2.205) and by taking into account that $R_1 = R_3$ and the inequality (2.204), we find

$$M_4^f \neq M_2^f. \tag{2.206}$$

By continuing the same line of reasoning, we can establish that

$$M_{2k+1}^f \neq M_{2k-1}^f, \tag{2.207}$$

and

$$M_{2k}^f \neq M_{2k-2}^f. \tag{2.208}$$

The last two inequalities show that some stabilization (accommodation) process is going on before a stable minor loop is asymptotically reached. This accommodation process is illustrated by Fig. 2.43.

We next proceed to the discussion of the identification problem. It turns out that the model (2.197)–(2.198) is very general in nature and it is not clear at this point how to solve the identification problem for this model. However, the identification problem becomes tractable if the dependence of μ-function on $f^{(m)}$ is factored out. In this way, we arrive at the following model:

$$f(t) = \iint_{R_{u(t)}} \nu(f^{(m)})\mu(\alpha, \beta)\hat{\gamma}_{\alpha\beta}u(t)\,d\alpha\,d\beta + r(u(t)), \tag{2.209}$$

where $f^{(m)}$ are defined in the same way as in (2.198), and the ν-function is assumed to be known. Thus, the identification problem consists in deter-

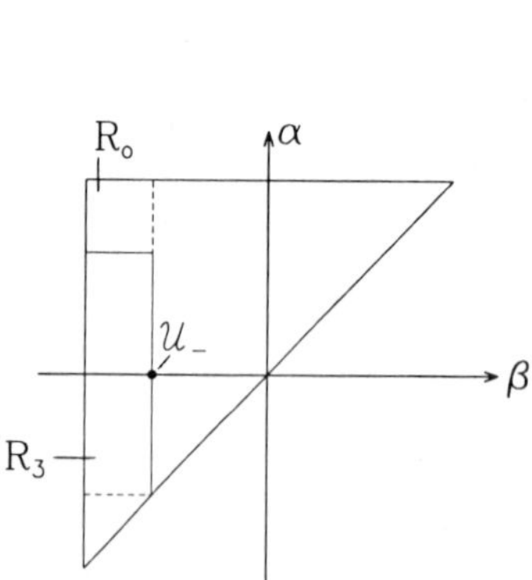

Fig. 2.42

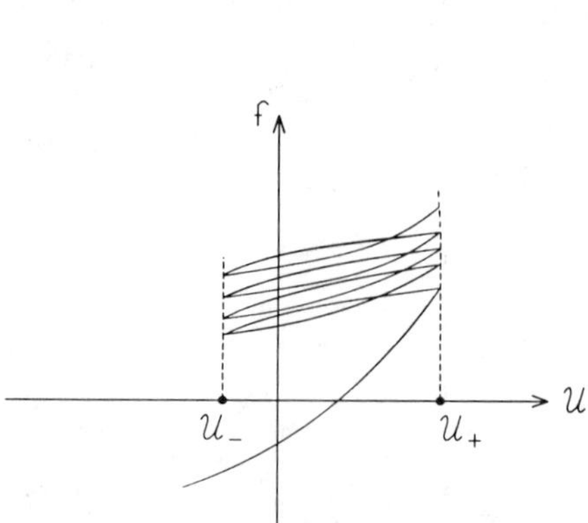

Fig. 2.43

mining the functions $r(u)$ and $\mu(\alpha, \beta)$ by fitting the model (2.209) to some experimental data. We begin with the determination of the function $r(u)$. Suppose that starting from the state of negative saturation the input is monotonically increased to some value u. As a result of this monotonic increase, the output will reach some value f_u^+ on the limiting ascending branch. According to (2.209), for this output value we have

$$f_u^+ = -\nu(f_-) \iint_{R(\alpha_0, \beta_0, u)} \mu(\alpha, \beta) d\alpha d\beta + r(u). \tag{2.210}$$

Now consider the state of positive saturation and suppose that the input is monotonically decreased to the same value u. In this case, the limiting descending branch will be followed and some output value f_u^- on this branch will be reached. According to (2.209), for the above-mentioned output value we find

$$f_u^- = \nu(f_+) \iint_{R(\alpha_0, \beta_0, u)} \mu(\alpha, \beta) d\alpha d\beta + r(u), \tag{2.211}$$

where as before f_+ is the output value in the state of positive saturation.

From (2.210) and (2.211), we derive

$$r(u) = \frac{\nu(f_+) f_u^+ + \nu(f_-) f_u^-}{\nu(f_+) + \nu(f_-)}. \tag{2.212}$$

Consider a particular case when the ν-function is even. In this case, since $f_- = -f_+$, we find

$$\nu(f_+) = \nu(f_-), \tag{2.213}$$

and from (2.112) we conclude

$$r(u) = \tfrac{1}{2}(f_u^+ + f_u^-). \tag{2.214}$$

This coincides with the previous result for the moving Preisach model.

We next turn to the determination of the μ-function in (2.209). Consider first-order reversal curves $f_{\alpha\beta}$ which are attached to the limiting ascending branch. By fitting these curves by the model (2.209), we arrive at the following equation:

$$f_{\alpha\beta} = \nu(f_\alpha) \iint_{R(\alpha, \beta_0, \beta, \beta)} \mu(\alpha', \beta') d\alpha' d\beta' - \nu(f_-) \iint_{R(\alpha_0, \beta_0, \alpha, \beta)} \mu(\alpha', \beta') d\alpha' d\beta' + r(\beta). \tag{2.215}$$

From (2.215), we obtain

$$\frac{f_{\alpha\beta} - r(\beta)}{\nu(f_\alpha)} = \iint_{R(\alpha, \beta_0, \beta, \beta)} \mu(\alpha', \beta') d\alpha' d\beta' - \frac{\nu(f_-)}{\nu(f_\alpha)} \iint_{R(\alpha_0, \beta_0, \alpha, \beta)} \mu(\alpha', \beta') d\alpha' d\beta'. \tag{2.216}$$

On the other hand, by fitting the limiting descending branch f_β^- by the model (2.209), we find

$$f_\beta^- = \nu(f_+) \iint\limits_{R(\alpha_0, \beta_0, \beta, \beta)} \mu(\alpha', \beta')d\alpha' d\beta' + r(\beta). \tag{2.217}$$

The expression (2.217) can be transformed as follows:

$$\frac{f_\beta^- - r(\beta)}{\nu(f_+)} = \iint\limits_{R(\alpha_0, \beta_0, \alpha, \beta)} \mu(\alpha', \beta')d\alpha' d\beta' + \iint\limits_{R(\alpha, \beta_0, \beta, \beta)} \mu(\alpha', \beta')d\alpha' d\beta'. \tag{2.218}$$

By introducing the function

$$T(\alpha, \beta) = \iint\limits_{R(\alpha_0, \beta_0, \alpha, \beta)} \mu(\alpha', \beta')d\alpha' d\beta' \tag{2.219}$$

and by using (2.216) and (2.219), we derive the following expression for $T(\alpha, \beta)$ in terms of experimentally measured data:

$$T(\alpha, \beta) = \frac{\nu(f_\alpha)[f_\beta^- - r(\beta)] - \nu(f_+)[f_{\alpha\beta} - r(\beta)]}{\nu(f_+)[\nu(f_\alpha) + \nu(f_-)]}. \tag{2.220}$$

By knowing the function $T(\alpha, \beta)$ and by employing the same reasoning as usual, we can derive from (2.219) the following expression for μ:

$$\mu(\alpha, \beta) = -\frac{\partial^2 T(\alpha, \beta)}{\partial\alpha\partial\beta}. \tag{2.221}$$

For the purpose of numerical implementation of the model (2.209) we do not need to know the function μ. This is because an explicit expression for the output in terms of the function $T(\alpha, \beta)$ can be derived. The derivation proceeds as follows. From (2.209) and Fig. 2.38, we find

$$f(t) = \sum_{k=1}^{n(t)} \nu(M_{2k-1}^f) \iint\limits_{R_{2k-1}} \mu(\alpha, \beta)d\alpha d\beta - \sum_{k=1}^{n(t)} \nu(M_{2k}^f) \iint\limits_{R_{2k}} \mu(\alpha, \beta)d\alpha d\beta. \tag{2.222}$$

Due to the definition of $T(\alpha, \beta)$, we have

$$\iint\limits_{R_{2k-1}} \mu(\alpha, \beta)d\alpha d\beta = T(M_{k+1}, m_k) - T(M_k, m_k). \tag{2.223}$$

Similarly, we find

$$\iint\limits_{R_{2k}} \mu(\alpha, \beta)d\alpha d\beta = T(M_{k+1}, m_{k+1}) - T(M_{k+1}, m_k), \tag{2.224}$$

where it is assumed that

$$T(M_1, m_0) = 0. \tag{2.225}$$

By substituting (2.223) and (2.224) into (2.222), we derive

$$f(t) = \sum_{k=1}^{n(t)} \nu(M^f_{2k-1})[T(M_{k+1}, m_k) - T(M_k, m_k)] - \sum_{k=0}^{n(t)} \nu(M^f_{2k})[T(M_{k+1}, m_{k+1}) - T(M_{k+1}, m_k)]. \tag{2.226}$$

This is the final expression for the output in terms of the function T which is directly related to the experimental data by the formula (2.220).

We will conclude this section by the discussion of one sufficient condition which guarantees the convergence of the stabilization (accommodation) process. This condition is given by the following inequality:

$$\max_{M^f} |\nu'(M^f)| \cdot \max_R \left| \iint_R \mu(\alpha, \beta) d\alpha d\beta \right| = q < 1, \tag{2.227}$$

where ν' stands for the derivative of the ν-function, and R is any rectangle within the limiting triangle.

The proof that the condition (2.227) guarantees the convergence of the stabilization process is given below for the simplest case. This is the same case which was discussed in the beginning of this section when we proved that the model (2.197) did account for accommodation. More general cases can be treated similarly.

From (2.209), we find

$$M^f_{2k+1} = -\nu(f_-) \iint_{R_0} \mu(\alpha, \beta) d\alpha d\beta - \nu(M^f_{2k}) \iint_{R_+} \mu(\alpha, \beta) d\alpha d\beta + r(u_+), \tag{2.228}$$

where the notation R_+ is used for R_{2k} because they do not change from one cycle to another.

Similarly, we have

$$M^f_{2k} = \nu(M^f_{2k-1}) \iint_{R_-} \mu(\alpha, \beta) d\alpha d\beta - \nu(f_-) \iint_{R_0} \mu(\alpha, \beta) d\alpha d\beta + r(u_-), \tag{2.229}$$

where R_- stands for R_{2k+1}.

From (2.228) and (2.229), we derive

$$M^f_{2k+1} - M^f_{2k-1} = -[\nu(M^f_{2k}) - \nu(M^f_{2k-2})] \iint_{R_+} \mu(\alpha, \beta) d\alpha d\beta, \tag{2.230}$$

$$M^f_{2k} - M^f_{2k-2} = [\nu(M^f_{2k-1}) - \nu(M^f_{2k-3})] \iint_{R_-} \mu(\alpha, \beta) d\alpha d\beta. \tag{2.231}$$

It is apparent that

$$|\nu(M_{2k}^f) - \nu(M_{2k-2}^f)| \le \max_{M^f} |\nu'(M^f)||M_{2k}^f - M_{2k-2}^f|, \tag{2.232}$$

$$|\nu(M_{2k-1}^f) - \nu(M_{2k-3}^f)| \le \max_{M^f} |\nu'(M^f)||M_{2k-1}^f - M_{2k-3}^f|. \tag{2.233}$$

From (2.227) and (2.230)–(2.233), we derive

$$|M_{2k+1}^f - M_{2k-1}^f| \le q|M_{2k}^f - M_{2k-2}^f|, \tag{2.234}$$

$$|M_{2k}^f - M_{2k-2}^f| \le q|M_{2k-1}^f - M_{2k-3}^f|. \tag{2.235}$$

By combining (2.234) and (2.235), we obtain

$$|M_{2k+1}^f - M_{2k-1}^f| \le q^2|M_{2k-1}^f - M_{2k-3}^f|, \tag{2.236}$$

$$|M_{2k+2}^f - M_{2k}^f| \le q^2|M_{2k}^f - M_{2k-2}^f|. \tag{2.237}$$

Inequalities (2.236) and (2.237) mean that extrema $\{M_{2k+1}^f\}$ and $\{M_{2k}^f\}$ form two contracting sequences. For this reason, there exist the limits:

$$\lim_{k\to\infty} M_{2k+1}^f = M_{u_+}^f, \tag{2.238}$$

$$\lim_{k\to\infty} M_{2k}^f = M_{u_-}^f. \tag{2.239}$$

Consequently, the condition (2.227) guarantees the convergence of the accommodation process.

6 Superconducting Hysteresis and Preisach-Type Models

The classical Preisach model has mostly been used for the description of magnetic hysteresis of ferromagnetic materials. However, it has been repeatedly emphasized in this book that the Preisach model is phenomenological in nature, and, for this reason, it can be employed for the description of hysteresis of any physical origin. To illustrate this statement, in this section we will apply the Preisach-type models to the description of superconducting hysteresis. This topic can also be of independent interest, because it is concerned with a nontraditional approach to studying superconducting hysteresis.

It is well known that high field (hard) type II superconductors exhibit magnetic hysteresis. The hysteresis in such superconductors occurs because the motion of flux filaments is pinned by defects such as voids, normal inclusions, dislocations, grain boundaries, compositional variations, etc. The pinning of flux filaments is instrumental in achieving high critical current densities. Thus, the very phenomenon that makes type II superconductors useful in practical applications is also responsible for magnetic hysteresis of these superconductors.

The critical state model has been proposed by C.P. Bean [49], [50] (see also [51]) to treat the magnetic hysteresis of hard superconductors. This model is based on the idealized (rectangular) electric field vs. current $\vec{E}(\vec{J})$ relation which implies that persistent currents up to a critical current density are always induced in superconductors. The Bean model has been further generalized [52] to take into account a dependence of critical current density on magnetic field. The critical state type models have also been tested experimentally and have proved to be fairly accurate for simple shape specimens (plane slabs, circular cross-section cylinders). It has as well been realized that the critical state type models have some intrinsic limitations. These limitations stem from idealizations of the $\vec{E}$ vs. $\vec{J}$ relation and also from the fact that the explicit solutions of the Maxwell equations even for idealized $\vec{E}$ vs. $\vec{J}$ relations are only possible for very simple specimen shapes.

In this section, it is shown that the critical state model and its generalizations are very particular cases of the classical Preisach model. On the basis of this fact, it is suggested to use the Preisach-type models for the description of superconducting hysteresis. In particular, it is proposed to evaluate hysteresis losses in superconductors by using the Preisach-type models and experimentally measured first- and second-order transition curves. It is important to keep in mind that the Preisach approach to the modelling of superconducting hysteresis is phenomenological in nature and does not intend to describe or explain actual physical causes of this hysteresis. This approach is, in a way, a sophisticated mathematical interpretation of experimental data. Its strength lies in direct connections with experimental data, and, for this reason, the Preisach approach can be of value in the design of superconducting devices.

First, we briefly describe some basic facts concerning the Bean critical state model of superconducting hysteresis. Consider a plane superconducting slab of thickness D (see Fig. 2.44) which is subject to an external time varying magnetic field $H_e(t)$. We will be interested in the B vs. H_e relation. Here B is an average magnetic flux density which is defined as

$$B = \frac{\mu_0}{D} \int_0^D H(x)dx, \tag{2.240}$$

where $H(x)$ is a magnetic field within the slab.

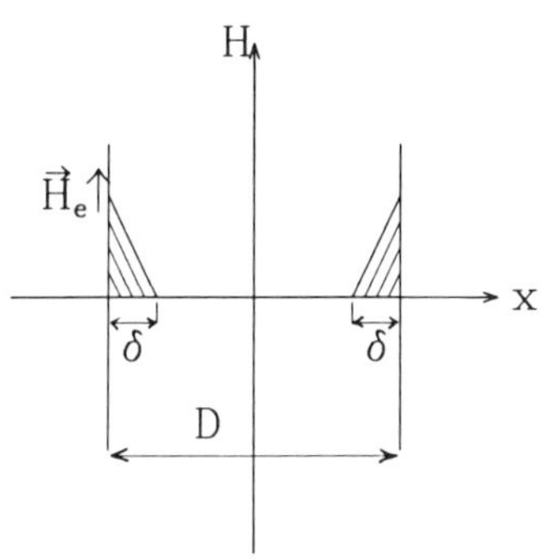

Fig. 2.44

In practice, B and H_e are quantities which are usually experimentally measured and it is their relation that exhibits hysteresis.

From (2.240) it follows that in order to compute B for any H_e, we have to find a magnetic field profile (magnetic field distribution) within the superconducting slab. The Bean model solves exactly this problem. The basic premise of the Bean model is that there exists a limiting (critical) superconducting current density J_c and that any voltage, however small, will induce this full current to flow. This premise is equivalent to nonlinear Ohm's law illustrated by Fig. 2.45. From the above premise also follows that current density profiles are rectangular in shape (Fig. 2.46).

We will next use the Ampère Law:

$$\operatorname{curl} \vec{H} = \vec{J}, \tag{2.241}$$

which for our problem can be written as

$$\frac{dH}{dx} = -J_c. \tag{2.242}$$

Since the critical current density J_c is constant, the expression (2.240) implies linear profiles of the magnetic field within the slab.

Having established the last fact, we next consider magnetic field distributions within the slab for different variations of the external magnetic field. Suppose that no magnetic field was present prior to the instant of time t_0. For times $t > t_0$, the external magnetic field H_e is monotonically increased until it reached some maximum value H_m. The monotonic increase of the external magnetic field induces persisting currents which result in the formation of linear profiles of the magnetic field shown in Fig. 2.44. It is easy to see from (2.242) that the depth of penetration of the magnetic field is given by

$$\delta = \frac{H_e}{J_c}. \tag{2.243}$$

It is also clear that

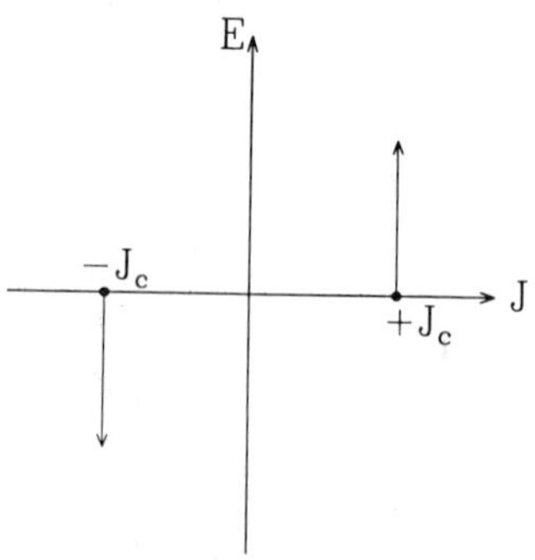

Fig. 2.45

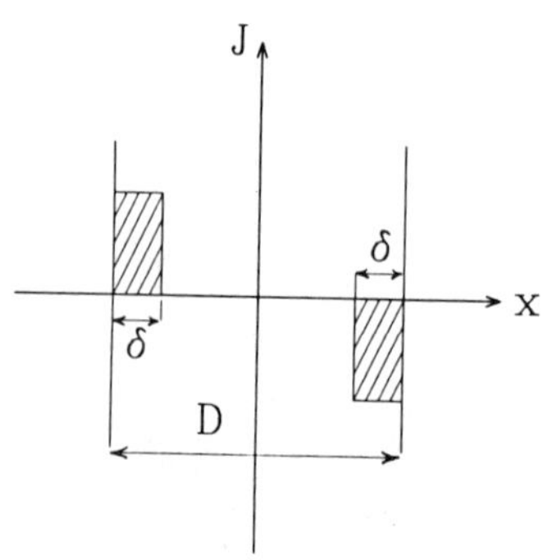

Fig. 2.46

$$\delta < \frac{D}{2}, \tag{2.244}$$

if

$$H_e < \frac{J_c D}{2} = H^*. \tag{2.245}$$

By using Fig. 2.44 and formulas (2.240) and (2.243), we find the average value of magnetic flux density

$$B = \frac{\mu_0 H_e \delta}{D} = \frac{\mu_0 H_e^2}{2H^*}. \tag{2.246}$$

Suppose now that after achieving the maximum value H_m, the external magnetic field is monotonically decreased to zero. As soon as the maximum value H_m is achieved, the motion of the previous linear profile is terminated and a new moving linear profile of magnetic field is formed. Due to the previously induced persisting currents, the previous profile stays still and is partially wiped out by the motion of new profile. The distribution of magnetic field within the slab at the instant of time when the external magnetic field is reduced to zero is shown in Fig. 2.47. This figure shows that there is non-zero (positive) average magnetic flux density which is given by

$$\tilde{B} = \frac{\mu_0 H_m^2}{4H^*} > 0. \tag{2.247}$$

This clearly suggests that the B vs. H_e relation exhibits hysteresis. We next demonstrate the validity of this statement by computing the hysteresis loop for the case of back-and-forth variation of the external magnetic field between $-H_0$ and $+H_0$. For the sake of simplicity of our computations, we shall assume that

$$H_0 \leq H^*.$$

We first consider the half-period when the external magnetic field is monotonically decreased. A typical magnetic field distribution for this half-period

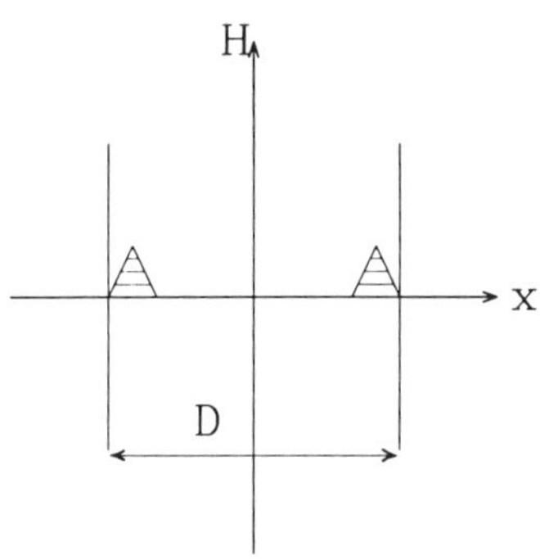

Fig. 2.47

is shown in Fig. 2.48. For the penetration depths δ_0 an δ, shown on this figure, we have

$$\delta_0 = \frac{H_0}{J_c}, \qquad \delta = \frac{H_0 - H_e}{2J_c}. \tag{2.248}$$

By using Fig. 2.48 and (2.248), we find the increment ΔB of the average magnetic flux density

$$\Delta B = \frac{2\mu_0}{D} \cdot \frac{(H_0 - H_e)\delta}{2} = \mu_0 \frac{(H_0 - H_e)^2}{4H^*}. \tag{2.249}$$

This leads to the following expression for the average magnetic flux density on the descending branch of hysteresis loop

$$B = B_0 - \Delta B = \frac{\mu_0 H_0^2}{2H^*} - \frac{\mu_0 (H_0 - H_e)^2}{4H^*}. \tag{2.250}$$

Consider now the half-period during which the external magnetic field is monotonically increased from $-H_0$ to $+H_0$. A typical magnetic field distribution for this half-period is shown in Fig. 2.49. By using this figure, as before we find

$$\Delta B = \frac{\mu_0 (H_0 + H_e)^2}{4H^*}, \tag{2.251}$$

and

$$B = -B_0 + \Delta B = -\frac{\mu_0 H_0^2}{2H^*} + \frac{\mu_0 (H_0 + H_e)^2}{4H^*}. \tag{2.252}$$

The expressions (2.250) and (2.252) can be combined into one formula

$$B = \pm \mu_0 \left[\frac{H_0^2}{2H^*} - \frac{(H_0 \mp H_e)^2}{4H^*} \right], \tag{2.253}$$

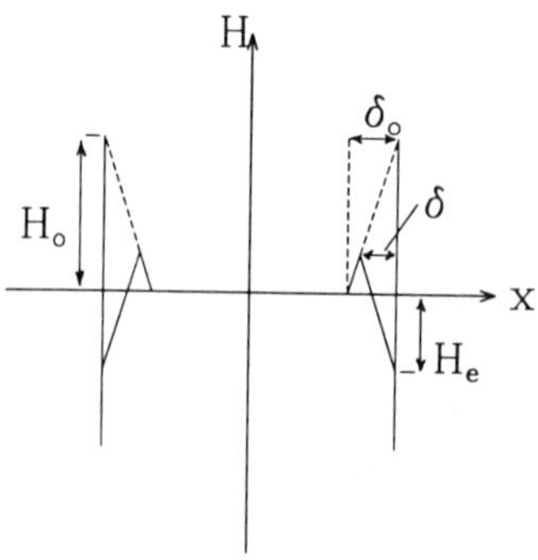

Fig. 2.48

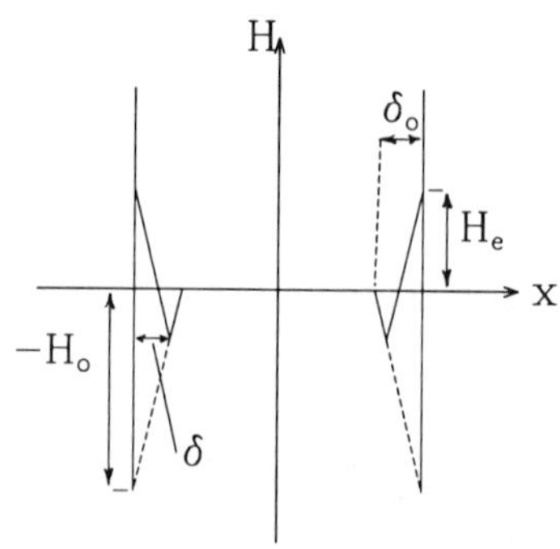

Fig. 2.49

where the upper signs correspond to the descending branch of the loop, while the lower signs correspond to the ascending branch.

On the basis of the previous discussion, the essence of the Bean model can now be summarized as follows. Each reversal of the magnetic field $H_e(t)$ on the boundary of superconducting slab results in the formation of a linear profile of the magnetic field. This profile penetrates the superconductor until another reversal value of the magnetic field on the boundary is reached. At this point, the motion of the previous profile is terminated and a new moving linear profile is formed. Due to the previously induced persisting currents, the previous linear profiles stay still and they represent past history which leaves its mark upon future values of average magnetic flux density. These persisting linear profiles of magnetic field may be partially or completely wiped out by new moving profiles.

It is clear from the above given description of the Bean model that the wiping-out property and congruency property hold for this model. Indeed, a moving linear profile will wipe out those persisting linear profiles if they correspond to the previous extremum values of $H_e(t)$ which are exceeded by a new extremum value. In this way, the effect of those previous extremum values of $H_e(t)$ on the future average values of magnetic flux density B will be completely eliminated. This means that the wiping-out property holds. It can also be shown that the congruency property of minor loops corresponding to the same reversal values of $H_e(t)$ holds as well. Indeed, consider two variations of external magnetic field $H_e^{(1)}(t)$ and $H_e^{(2)}(t)$. Suppose that these external fields may have different past histories, but starting from some instant of time t_0 they vary back-and-forth between the same reversal values. It is apparent from the above description of the Bean model that these back-and-forth variations will affect in the *same* way *equal* surface layers of superconductors. Consequently, these variations will result in equal increments of B, which is tantamount to the congruency of the corresponding minor loops.

In the case of generalized critical state models [52], the linear profiles of the magnetic field within superconductors are replaced by curved profiles. However, the creation and motion of these profiles are basically governed by the same rules as in the case of the Bean model. As a result, the previous reasoning holds, and, consequently, the wiping-out property and the congruency property are valid for the generalized critical state model as well. It has been established in Chapter I that the wiping-out property and congruency property constitute necessary and sufficient conditions for the representation of actual hysteresis nonlinearity by the classical Preisach model. Thus, we conclude that the Bean model and generalized critical state models are particular cases of the classical Preisach model.

It is instructive to find such a function $\mu(\alpha, \beta)$ for which the Preisach model coincides with the Bean model. To do this, consider a "major" loop formed when the external magnetic field varies back-and-forth between $+H_0$ and $-H_0$. Consider first-order transition curves $B_{\alpha\beta}$ attached to the ascending branch of the above-mentioned loop. The curves $B_{\alpha\beta}$ are formed when, after

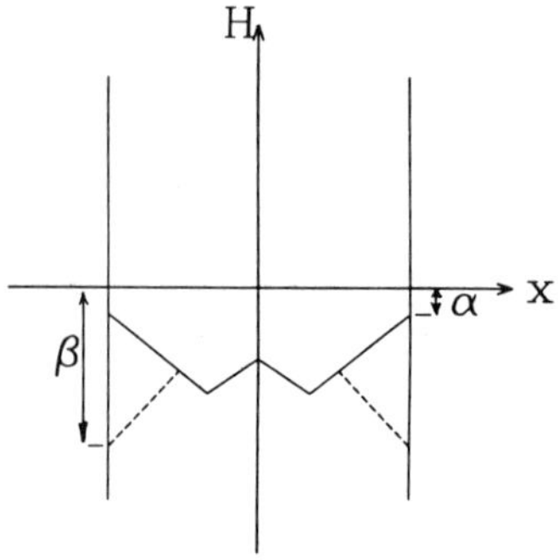

Fig. 2.50

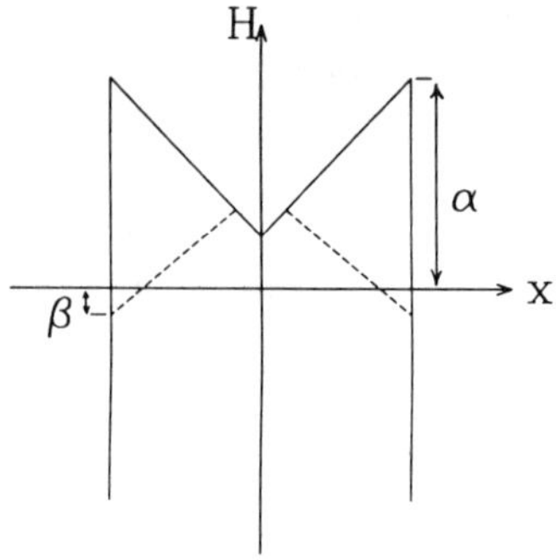

Fig. 2.51

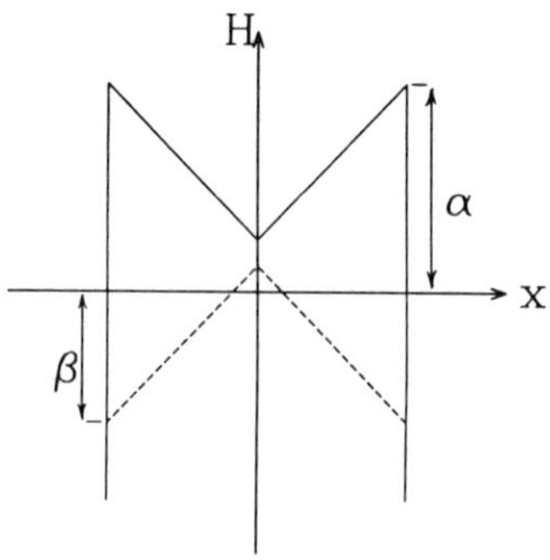

Fig. 2.52

reaching the value $-H_0$, the external magnetic field is monotonically increased to the value α and subsequently monotonically decreased to the value β. Depending on particular values of α and β, we may have three typical field distributions shown in Fig. 2.50, 2.51, and 2.52. We will use these figures to evaluate the function

$$F(\alpha, \beta) = \tfrac{1}{2}(B_\alpha - B_{\alpha\beta}). \tag{2.254}$$

Figure 2.50 is valid under the condition

$$H_0 + \alpha \leq 2H^*. \tag{2.255}$$

From this figure we find

$$F(\alpha, \beta) = \frac{\mu_0(\alpha - \beta)^2}{8H^*}. \tag{2.256}$$

Figure 2.51 holds when

$$H_0 + \alpha \geq 2H^*, \qquad \alpha - \beta \leq 2H^*. \tag{2.257}$$

By using this figure, we derive

$$F(\alpha, \beta) = \frac{\mu_0(\alpha - \beta)^2}{8H^*}. \tag{2.258}$$

Finally, the distribution of magnetic field shown in Fig. 2.52 occurs when

$$H_0 + \alpha \geq 2H^* \quad \text{and} \quad \alpha - \beta \geq 2H^*. \tag{2.259}$$

From Fig. 2.52, we obtain

$$F(\alpha, \beta) = \frac{\mu_0}{2}(\alpha - \beta - H^*). \tag{2.260}$$

The expressions (2.256), (2.258), and (2.260) can be combined into one formula:

$$F(\alpha, \beta) = \begin{cases} \dfrac{\mu_0(\alpha - \beta)^2}{8H^*} & \text{if } 0 < \alpha - \beta \leq 2H^*, |\alpha| \leq H_0, |\beta| \leq H_0, \\ \dfrac{\mu_0}{2}(\alpha - \beta - H^*), & \text{if } \alpha - \beta \geq 2H^*, |\alpha| \leq H_0, |\beta| \leq H_0. \end{cases} \tag{2.261}$$

By using (2.261) and the in Chapter I derived formula

$$\mu(\alpha, \beta) = -\frac{\partial^2 F(\alpha, \beta)}{\partial\alpha\partial\beta}, \tag{2.262}$$

we find

$$\mu(\alpha, \beta) = \begin{cases} \dfrac{\mu_0}{4H^*}, & \text{if } 0 < \alpha - \beta \leq 2H^*, |\alpha| \leq H_0, |\beta| \leq H_0, \\ 0, & \text{otherwise.} \end{cases} \tag{2.263}$$

The support of $\mu(\alpha, \beta)$ given by (2.263) is illustrated in Fig. 2.53.

The facts that the classical Preisach model contains the Bean model and its generalizations as very particular cases and that the critical state type models have been in fairly good agreement with many experiments suggest that the Preisach-type models may be quite accurate in the description of superconducting hysteresis. This statement is also supported by the fact that the

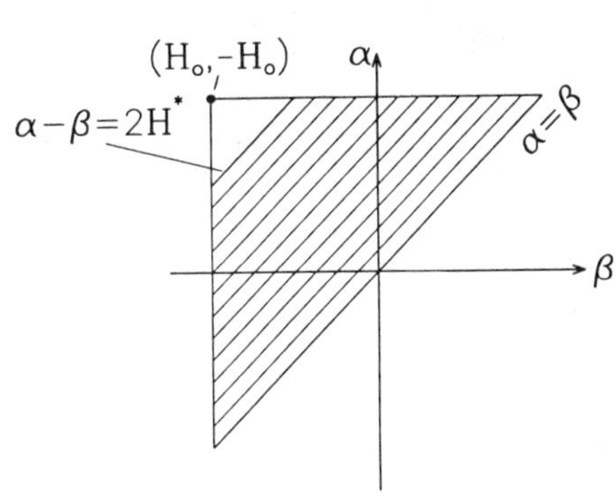

Fig. 2.53

Preisach-type models incorporate a much larger volume of experimental data than the critical state type models. Indeed, the identification of Preisach-type models requires sets of experimentally measured first- and second-order transition curves, while the critical state type models use only few parameters (critical current density, slab thickness, etc.).

One of the main gains in using the Preisach-type models for the description of superconducting hysteresis is that these models allow one to evaluate hysteretic losses for arbitrary (not necessarily cyclic) time variations of the external magnetic field. For instance, if one is interested in hysteretic losses occurring in superconductors during the monotonic increase of the external magnetic field from H_- to H_+, then the formula (1.115) can be employed. For our problem, this formula can be represented as

$$Q(H_+, H_-) = -\frac{1}{2}\left[\int_{H_-}^{H_+} F(\alpha, H_-)d\alpha + \int_{H_-}^{H_+} F(H_+, \beta)d\beta - (H_+ - H_-)F(H_+, H_-)\right], \tag{2.264}$$

where $Q(H_+, H_-)$ is a hysteretic loss and the function F is defined by (2.254).

The last formula evaluates the hysteretic losses if the classical Preisach model is used for the description of superconducting hysteresis. If the nonlinear Preisach model is employed for the above purpose, then the hysteretic losses can be evaluated by using the formula (2.77). For our problem, this formula can be represented in the form

$$Q(H_-, H_+) = \frac{1}{2}\int_{H_-}^{H_+} P(H_+, H_-, H_e)dH_e, \tag{2.265}$$

where the function P is related to the second order transition curves by the expression (see 2.32)

$$P(\alpha, \beta, H) = B_{\alpha H} - B_{\alpha\beta H}. \tag{2.266}$$

Formulas (2.264) and (2.266) express noncyclic hysteretic losses directly in terms of experimentally measured first- and second-order transition curves. This suggests that these transition curves may emerge as standard experimental data for the evaluation of hysteretic losses in superconductors. It is also worthwhile to keep in mind that noncyclic hysteretic losses may be of practical significance in many applications, for instance, when one is interested in losses occurring in the process of loading (or unloading) superconducting magnets.

7 Magnetostrictive Hysteresis and Preisach Models with Two Inputs

The essence of magnetostrictive phenomenon is the dependence of strain on magnetization. Since the magnetization can be varied by applied magnetic fields, this opens the opportunity to control the strain of magnetostrictive rods by magnetic fields. For this reason, magnetostrictive materials (especially Terfenol type materials with gigantic magnetostriction) are very attractive as actuator materials for many applications such as robotics, active vibration damping, micromotors, etc. However magnetostrictive materials exhibit hysteresis, which represents a problem for fine positioning applications. If hysteretic effects of magnetostrictive materials could be predicted, then actuator controllers could be designed to correct for these effects. This would result in high precision actuators powered by magnetostrictive materials. It is clear that mathematical models for magnetostrictive hysteresis could facilitate the design of the above controllers.

It is known that magnetostrictive materials exhibit hysteretic behavior with respect to variations of two variables: magnetic field and stress. Thus the problem of developing the Preisach type models with two inputs presents itself. These models are discussed in this section. It is quite possible that they may find some applications beyond the area of magnetostrictive hysteresis.

For the sake of generality, we consider a hysteresis nonlinearity which can be characterized by two inputs $u(t)$ and $v(t)$ and an output $f(t)$. In magnetostriction applications $u(t)$ is the magnetic field, $v(t)$ is the stress, while $f(t)$ is the strain. We shall discuss the following Preisach type model with two inputs:

$$f(t) = \iint\limits_{\alpha \geq \beta} \mu(\alpha, \beta, v(t))\hat{\gamma}_{\alpha\beta}u(t)d\alpha d\beta + \iint\limits_{\alpha \geq \beta} \nu(\alpha, \beta, u(t))\hat{\gamma}_{\alpha\beta}v(t)d\alpha d\beta. \quad (2.267)$$

In the above model, the dependence of the functions μ and ν on $v(t)$ and $u(t)$, respectively, reflects the cross-coupling between two inputs.

The investigation of the model (2.267) is greatly facilitated by its equivalent representation as a model with moving supports for $\mu(\alpha, \beta, v(t))$ and $\nu(\alpha, \beta, u(t))$. To arrive at this representation, we introduce the sets $S^+_{u(t)}$, $R_{u(t)}$, $S^-_{u(t)}$, $S^+_{v(t)}$, $R_{v(t)}$ and $S^-_{v(t)}$, which can be defined by the same formulas as (2.2)–(2.4). By using these sets, the expression (2.267) can be transformed as follows:

$$\begin{aligned} f(t) = &\iint\limits_{R_{u(t)}} \mu(\alpha, \beta, v(t))\hat{\gamma}_{\alpha\beta}u(t)d\alpha d\beta + \iint\limits_{R_{v(t)}} \nu(\alpha, \beta, u(t))\hat{\gamma}_{\alpha\beta}v(t)d\alpha d\beta \\ &+ \iint\limits_{S^+_{u(t)}} \mu(\alpha, \beta, v(t))d\alpha d\beta - \iint\limits_{S^-_{u(t)}} \mu(\alpha, \beta, v(t))d\alpha d\beta \\ &+ \iint\limits_{S^+_{v(t)}} \nu(\alpha, \beta, u(t))d\alpha d\beta - \iint\limits_{S^-_{v(t)}} \nu(\alpha, \beta, u(t))d\alpha d\beta. \end{aligned} \quad (2.268)$$

We next find a simple expression for the last four terms in (2.268). Consider first a state of negative saturation and a subsequent monotonic increase of inputs until they reach some values $u(t)$ and $v(t)$. Let $f^+_{u(t)v(t)}$ be the resulting output value. Then, according to (2.268), we find:

$$f^{+}_{u(t)v(t)} = -\iint\limits_{R_{u(t)}} \mu(\alpha, \beta, v(t))d\alpha d\beta - \iint\limits_{R_{v(t)}} \nu(\alpha, \beta, u(t))d\alpha d\beta$$

$$+ \iint\limits_{S^{+}_{u(t)}} \mu(\alpha, \beta, v(t))d\alpha d\beta - \iint\limits_{S^{-}_{u(t)}} \mu(\alpha, \beta, v(t))d\alpha d\beta$$

$$+ \iint\limits_{S^{+}_{v(t)}} \nu(\alpha, \beta, u(t))d\alpha d\beta - \iint\limits_{S^{-}_{v(t)}} \nu(\alpha, \beta, u(t))d\alpha d\beta. \tag{2.269}$$

Similarly, starting from the state of positive saturation, we can define $f^{-}_{u(t)v(t)}$ and derive the following expression:

$$f^{-}_{u(t)v(t)} = +\iint\limits_{R_{u(t)}} \mu(\alpha, \beta, v(t))d\alpha d\beta + \iint\limits_{R_{v(t)}} \nu(\alpha, \beta, u(t))d\alpha d\beta$$

$$+ \iint\limits_{S^{+}_{u(t)}} \mu(\alpha, \beta, v(t))d\alpha d\beta - \iint\limits_{S^{-}_{u(t)}} \mu(\alpha, \beta, v(t))d\alpha d\beta$$

$$+ \iint\limits_{S^{+}_{v(t)}} \nu(\alpha, \beta, u(t))d\alpha d\beta - \iint\limits_{S^{-}_{v(t)}} \nu(\alpha, \beta, u(t))d\alpha d\beta. \tag{2.270}$$

From (2.269) and (2.270) we derive:

$$\frac{1}{2}(f^{+}_{u(t)v(t)} + f^{-}_{u(t)v(t)}) = \iint\limits_{S^{+}_{u(t)}} \mu(\alpha, \beta, v(t))d\alpha d\beta - \iint\limits_{S^{-}_{u(t)}} \mu(\alpha, \beta, v(t))d\alpha d\beta$$

$$+ \iint\limits_{S^{+}_{v(t)}} \nu(\alpha, \beta, u(t))d\alpha d\beta - \iint\limits_{S^{-}_{v(t)}} \nu(\alpha, \beta, u(t))d\alpha d\beta. \tag{2.271}$$

By substituting (2.271) into (2.268), we arrive at the following representation of the model (2.267):

$$f(t) = \iint\limits_{R_{u(t)}} \mu(\alpha, \beta, v(t))\hat{\gamma}_{\alpha\beta}u(t)d\alpha d\beta + \iint\limits_{R_{v(t)}} \nu(\alpha, \beta, u(t))\hat{\gamma}_{\alpha\beta}v(t)d\alpha d\beta$$

$$+ \frac{1}{2}(f^{+}_{u(t)v(t)} + f^{-}_{u(t)v(t)}). \tag{2.272}$$

The last expression can be regarded as a generalization of (2.12).

By using the same line of reasoning as in previous sections, it can be established that the model (2.272) has the following two characteristic properties.

Wiping-Out Property

Only the alternating series of past dominant extrema of $u(t)$ and $v(t)$ are stored by the model (2.272), while all other past extrema of $u(t)$ and $v(t)$ are wiped out.

Property of Equal Vertical Chords

All minor hysteretic loops corresponding to the same consecutive extremum values of $u(t)$ for the same fixed value of v have equal vertical chords regardless of the past history of variations of $u(t)$ and $v(t)$. The same is true for minor hysteretic loops formed as a result of back-and-forth variations of $v(t)$ for any fixed value of u.

It turns out that the model (2.272) has also the following distinct property.

Path Independence Property

Consider two points (u_1, v_1) and (u_2, v_2) on the $u - v$ plane and a set of paths connecting these points and corresponding to monotonic *variations of* both *$v(t)$ and $v(t)$ (see Figure 2.54). Then, the output increment predicted by the model (2.272) does not depend on a particular monotonic path between the points (u_1, v_1) and (u_2, v_2).*

Proof

For any instant of time the model (2.272) can be represented as

$$f(t) = \iint_{S^+(t)} \mu(\alpha, \beta, v(t))d\alpha d\beta - \iint_{S^-(t)} \mu(\alpha, \beta, v(t))d\alpha d\beta$$
$$+ \iint_{\Omega^+(t)} \nu(\alpha, \beta, u(t))d\alpha d\beta - \iint_{\Omega^-(t)} \nu(\alpha, \beta, u(t))d\alpha d\beta$$
$$+ \frac{1}{2}(f^+_{u(t)v(t)} + f^-_{u(t)v(t)}). \tag{2.273}$$

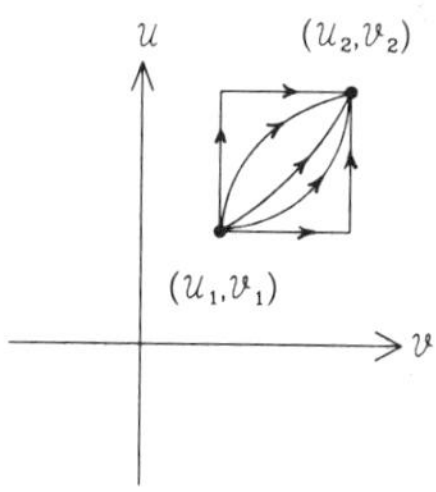

Fig. 2.54

Here: $S^+(t)$, $S^-(t)$, $\Omega^+(t)$ and $\Omega^-(t)$ are positive and negative sets on which $R_{u(t)}$ and $R_{v(t)}$ are subdivided, respectively.

It is clear that the above subdivision on positive and negative sets will be the same at the point (u_2, v_2) for all monotonic paths connecting this point with (u_1, v_1). This is because the monotonic variations of $u(t)$ and $v(t)$ will result in the same modification of negative and positive sets upon achieving the values u_2 and v_2, respectively. From here it is easy to conclude the validity of the above property.

We next proceed to the discussion of the identification problem for the model (2.272). In order to determine the functions $\mu(\alpha, \beta, v)$ and $\nu(\alpha, \beta, u)$ we shall use the first-order transition curves $f_{\alpha\beta v}$ and $f_{u\alpha\beta}$. As before, these curves are measured for piece-wise monotonic input variations started from the state of negative saturation. Namely, $f_{\alpha\beta v}$ is the output resulting from monotonic increases of two inputs to the values α and v, respectively, and subsequent monotonic decrease of $u(t)$ to the value β. The output value $f_{u\alpha\beta}$ can be determined in a similar way.

We shall define the functions

$$F(\alpha, \beta, v) = \tfrac{1}{2}(f_{\alpha\beta v} - f^{+}_{\beta v}), \tag{2.274}$$

$$G(\alpha, \beta, u) = \tfrac{1}{2}(f_{u\alpha\beta} - f^{+}_{u\beta}). \tag{2.275}$$

By using (2.272) and the $\alpha - \beta$ diagram technique, it is easy to show that

$$F(\alpha, \beta, v) = \iint\limits_{R(\alpha,\beta)} \mu(\alpha', \beta', v)\,d\alpha'\,d\beta' = \int_{\beta}^{\alpha}\left(\int_{\beta_0}^{\beta} \mu(\alpha', \beta', v)\,d\beta'\right)d\alpha', \tag{2.276}$$

$$G(\alpha, \beta, u) = \iint\limits_{R(\alpha,\beta)} \nu(\alpha', \beta', u)\,d\alpha'\,d\beta' = \int_{\beta}^{\alpha}\left(\int_{\beta_0}^{\beta} \nu(\alpha', \beta', u)\,d\beta'\right)d\alpha', \tag{2.277}$$

where $R(\alpha, \beta)$ is the rectangle shown in Figure 2.55.

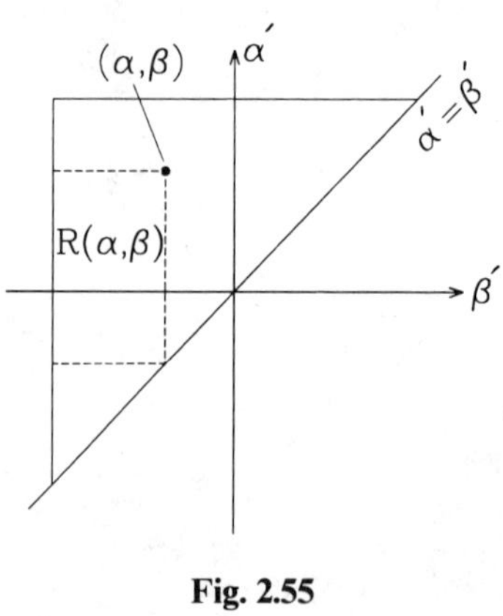

Fig. 2.55

From (2.276) and (2.277) we derive

$$\mu(\alpha, \beta, v) = \frac{\partial^2 F(\alpha, \beta, v)}{\partial\alpha\partial\beta}, \tag{2.278}$$

$$\nu(\alpha, \beta, u) = \frac{\partial^2 G(\alpha, \beta, u)}{\partial\alpha\partial\beta}. \tag{2.279}$$

As far as the numerical implementation of the model (2.272) is concerned, it can be accomplished without the formulas (2.278) and (2.279) but by employing explicit expressions for the integrals in (2.272) in terms of the functions $F(\alpha, \beta, v)$ and $G(\alpha, \beta, u)$. The derivation of these expressions proceeds as follows.

For the first integral in (2.272) we find:

$$\iint\limits_{R_{u(t)}} \mu(\alpha, \beta, v(t))\hat{\gamma}_{\alpha\beta}u(t)d\alpha d\beta = 2\iint\limits_{S^+(t)} \mu(\alpha, \beta, v(t))d\alpha d\beta - \iint\limits_{R_{u(t)}} \mu(\alpha, \beta, v(t))d\alpha d\beta. \tag{2.280}$$

Similarly

$$\iint\limits_{R_{v(t)}} \nu(\alpha, \beta, u(t))\hat{\gamma}_{\alpha\beta}v(t)d\alpha d\beta = 2\iint\limits_{\Omega^+(t)} \mu(\alpha, \beta, u(t))d\alpha d\beta - \iint\limits_{R_{v(t)}} \mu(\alpha, \beta, u(t))d\alpha d\beta. \tag{2.281}$$

From (2.269) and (2.270), we derive

$$\frac{1}{2}(f^{+}_{u(t)v(t)} - f^{-}_{u(t)v(t)}) = -\iint\limits_{R_{u(t)}} \mu(\alpha, \beta, v(t))d\alpha d\beta - \iint\limits_{R_{v(t)}} \mu(\alpha, \beta, u(t))d\alpha d\beta. \tag{2.282}$$

By substituting (2.280) and (2.281) into (2.272) and taking (2.282) into account, we obtain:

$$f(t) = 2\left[\iint\limits_{S^+(t)} \mu(\alpha, \beta, v(t))d\alpha d\beta + \iint\limits_{\Omega^+(t)} \nu(\alpha, \beta, u(t))d\alpha d\beta\right] + f^{+}_{u(t)v(t)} \tag{2.283}$$

A typical geometry for the set $S^+(t)$ is shown in Figure 2.56. From this figure follows

$$\iint\limits_{S^+(t)} \mu(\alpha, \beta, v(t))d\alpha d\beta = \sum_{k=1}^{n(t)} \iint\limits_{R_k} \mu(\alpha, \beta, v(t))d\alpha d\beta. \tag{2.284}$$

By using (2.276), we find

$$\iint\limits_{R_k} \mu(\alpha, \beta, v(t))d\alpha d\beta = F(M_k^{(u)}, m_k^{(u)}, v(t)) - F(M_{k+1}^{(u)}, m_k^{(u)}, v(t)) \tag{2.285}$$

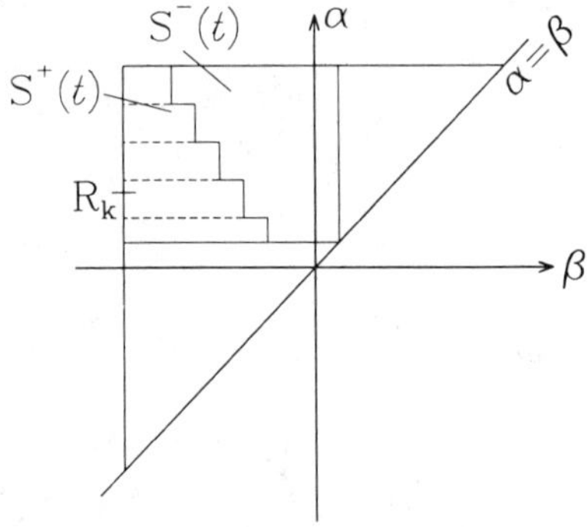

Fig. 2.56

where $M_k^{(u)}$ and $m_k^{(u)}$ form the alternating series of past dominant extrema of $u(t)$. From (2.284) and (2.285) we conclude

$$\iint_{S^+(t)} \mu(\alpha, \beta, v(t))d\alpha d\beta = \sum_{k=1}^{n(t)} [F(M_k^{(u)}, m_k^{(u)}, v(t)) - F(M_{k+1}^{(u)}, m_k^{(u)}, v(t))]. \quad (2.286)$$

By using the same line of reasoning as before, we can derive the expression

$$\iint_{\Omega^+(t)} \nu(\alpha, \beta, u(t))d\alpha d\beta = \sum_{k=1}^{l(t)} [G(M_k^{(v)}, m_k^{(v)}, u(t)) - G(M_{k+1}^{(v)}, m_k^{(v)}, u(t))]. \quad (2.287)$$

By substituting (2.287) and (2.286) into (2.283), we obtain

$$f(t) = f_{u(t)v(t)}^{+} + 2 \sum_{k=1}^{n(t)} [F(M_k^{(u)}, m_k^{(u)}, v(t)) - F(M_{k+1}^{(u)}, m_k^{(u)}, v(t))]$$
$$+ 2 \sum_{k=1}^{l(t)} [G(M_k^{(v)}, m_k^{(v)}, u(t)) - G(M_{k+1}^{(v)}, m_k^{(v)}, u(t))]. \quad (2.288)$$

This is the final expression for the output in terms of experimentally measured functions F and G defined by (2.274) and (2.275), respectively.

We next formulate the following result.

Representation Theorem

The wiping-out property, the property of equal vertical chords and the path independence property constitute necessary and sufficient conditions for the representation of actual two input hysteresis nonlinearities by the model (2.272) on the set of piece-wise monotonic inputs.

The proof of this theorem closely parallels the proof of the representation theorem from Section 2 of this chapter. The only new element is the utilization of the path independence property. By using this property, it can be easily

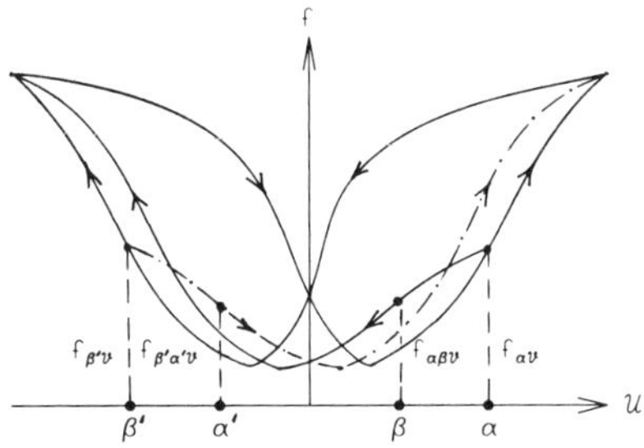

Fig. 2.57

shown that an arbitrary simultaneous piece-wise monotonic variation of $u(t)$ and $v(t)$ can be reduced to two consecutive piece-wise monotonic *single* input variations during which the other input remains constant. Further details of the proof of the theorem are left to the reader.

In conclusion of this section, we consider one peculiar property of strain hysteresis in comparison with magnetic hysteresis. This peculiarity stems from even symmetry of strain hysteresis which exhibits butterfly-shaped major hysteretic loops with respect to variations of magnetic field (see Fig. 2.57). It is clear that first-order transition curves $f_{\alpha\beta v}$ will also exhibit even symmetry. This can be expressed mathematically as

$$f_{\beta'\alpha' v} = f_{\alpha\beta v} \quad \text{and} \quad f^{-}_{\alpha' v} = f^{+}_{\beta v}, \tag{2.289}$$

if

$$\beta' = -\alpha \quad \text{and} \quad \alpha' = -\beta. \tag{2.290}$$

By using (2.289) and (2.290), it can be easily proven that

$$\mu(\alpha, \beta, v) = -\mu(-\beta, -\alpha, v). \tag{2.291}$$

Thus the function $\mu(\alpha, \beta, v)$ is of odd symmetry with respect to the line $\alpha = -\beta$. In the case of magnetic hysteresis, the μ-function has an even symmetry with respect to the same line.

It follows from (2.291) that the function $\mu(\alpha, \beta, v)$ can not be positive everywhere on the $\alpha - \beta$ plane. Actually, this function changes its sign across the line $\alpha = -\beta$. As a result, first-order and higher-order transition curves are not enclosed by major hysteretic loops. This property has been observed and it also distinguishes the strain hysteresis from magnetic hysteresis.

8 Experimental Testing of Preisach-Type Models of Hysteresis

In our previous discussions of Preisach-type models of hysteresis we consistently tried to establish the necessary and sufficient conditions under which actual hysteresis nonlinearities can be represented by these models. These conditions establish the limits of exact applicability of Preisach-type models. As a result, these conditions allow one to judge to what extent one or another Preisach-type model is general. These conditions also clearly reveal the phenomenological nature of Preisach-type models. However, it is unrealistic to expect that actual hysteresis nonlinearities will satisfy these conditions exactly. For instance, the experimental results presented in [53] show that neither the congruency property nor the property of equal vertical chords of comparable minor loops are satisfied for some typical materials used in magnetic recording. For this reason, the classical and nonlinear Preisach models as well as some other Preisach-type models can not be expected to be absolutely accurate. However, we may try to use these models as approximate ones. Then the question arises how to judge the accuracy of these models.

It has been emphasized in this book that branching is the essence of hysteresis. Consequently, any mathematical model of hysteresis is expected to predict this branching. In other words, models of hysteresis are expected to predict higher-order reversal curves (branches) which occur after several (or many) reversals of input. This prediction should be based on some limited experimental data used for the identification of hysteresis models. Thus, the accuracy of the classical, nonlinear and restricted Preisach models can be measured by their abilities to predict higher-order reversal curves on the basis of information provided by the first- and second-order reversal curves. This is exactly the criterion which has been chosen for the experimental testing of accuracy of the Preisach-type models.

This testing has been performed for two typical particulate materials: Co-coated $\gamma - Fe_2O_3$ and iron magnetic tape materials. The experimental work consisted of two main stages. During the first stage, the first- and second-order transition curves were measured by using a vibrating sample magnetometer. Experimentally measured first-order transition curves for the above two materials are shown in Fig. 2.58 and 2.59, respectively. For each first-order transition curve there is a set of second-order transition curves which are attached to the above first-order curve. Experimentally measured sets of second-order transition curves are shown in Figs. 2.60 and 2.61, respectively. These first- and second-order transition curves were used for the identification of the classical, nonlinear, and restricted Preisach models.

During the second stage of the performed experiments, several sets of higher-order (with the order up to six) transition curves were measured. The sequences of field reversals for these higher-order transition curves were chosen in such a way that these higher-order transition curves were more or less uniformly (and densely) distributed over the area enclosed by major hysteresis loops. This was important in order to get the information about *uniform* accuracy of Preisach-type models. Then, by using the developed computer software which implements the Preisach-type models, various sets of higher-order transition curves predicted by the Preisach-type models were computed for the same sequences of field reversals as in the experiment. The comparison of the higher-order transition curves predicted by the classical

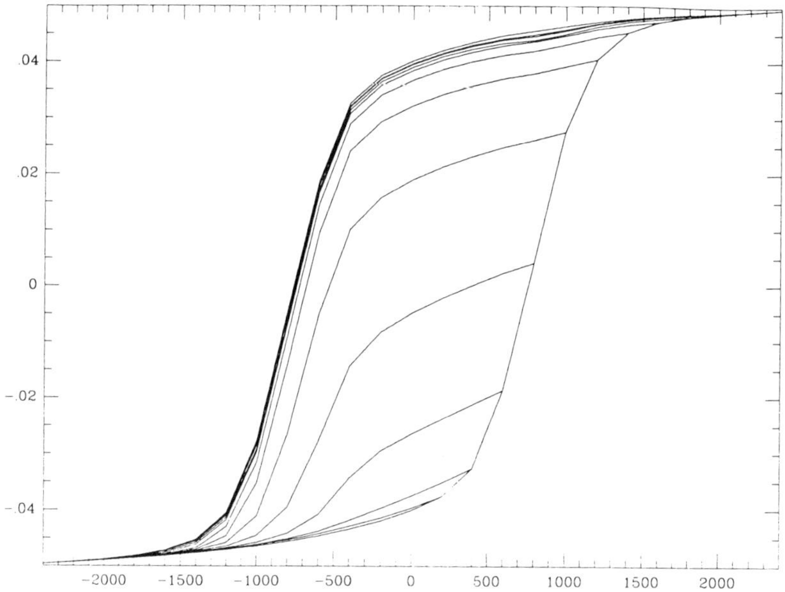

Fig. 2.58

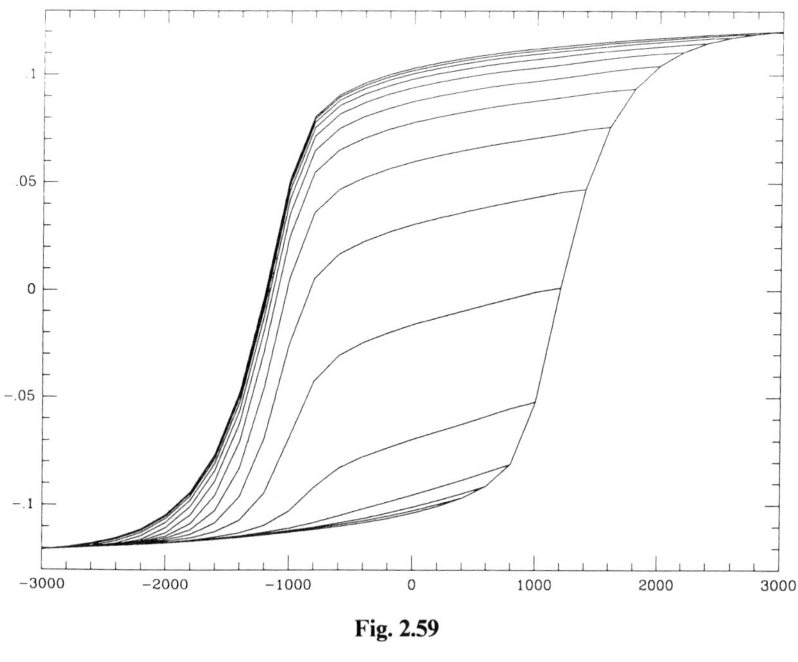

Fig. 2.59

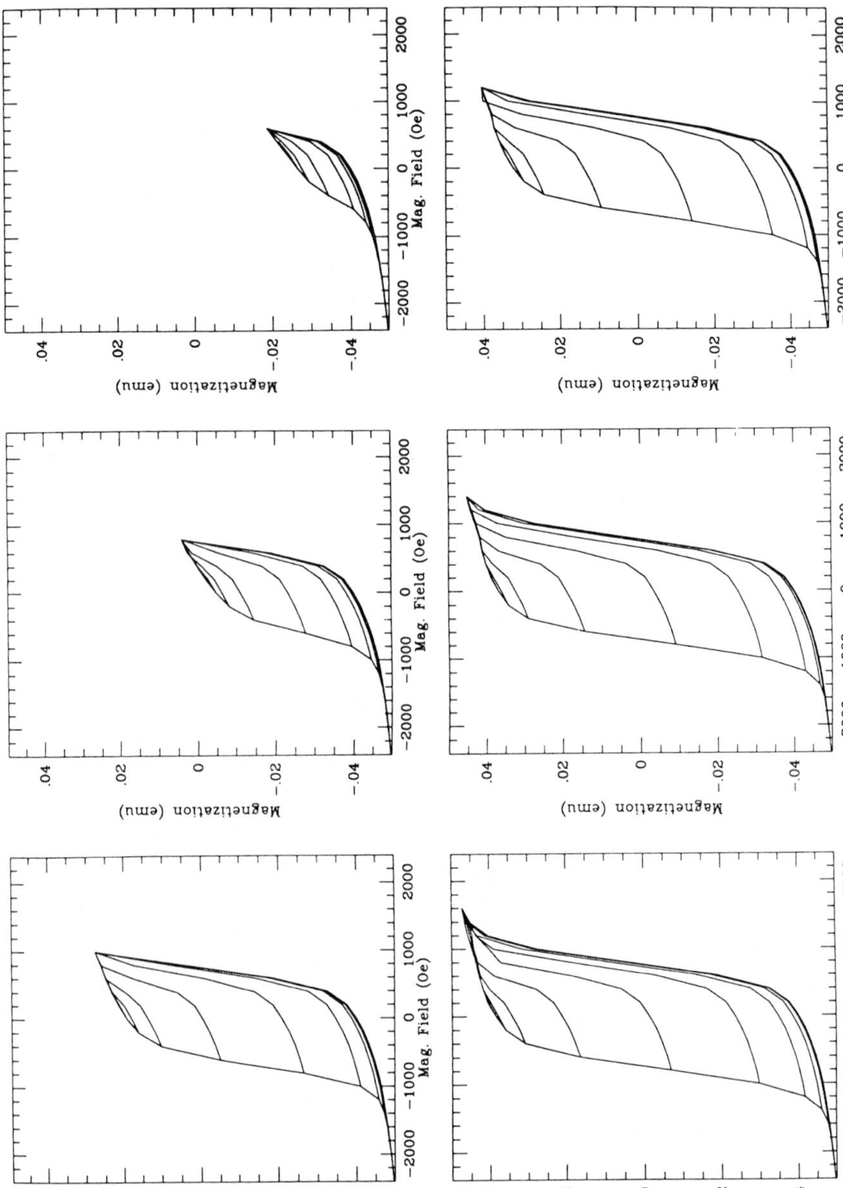

Magnetization (emu)
Mag. Field (Oe)
.04
.02
0
-.02
-.04
-2000
-1000
0
1000
2000

Fig. 2.60

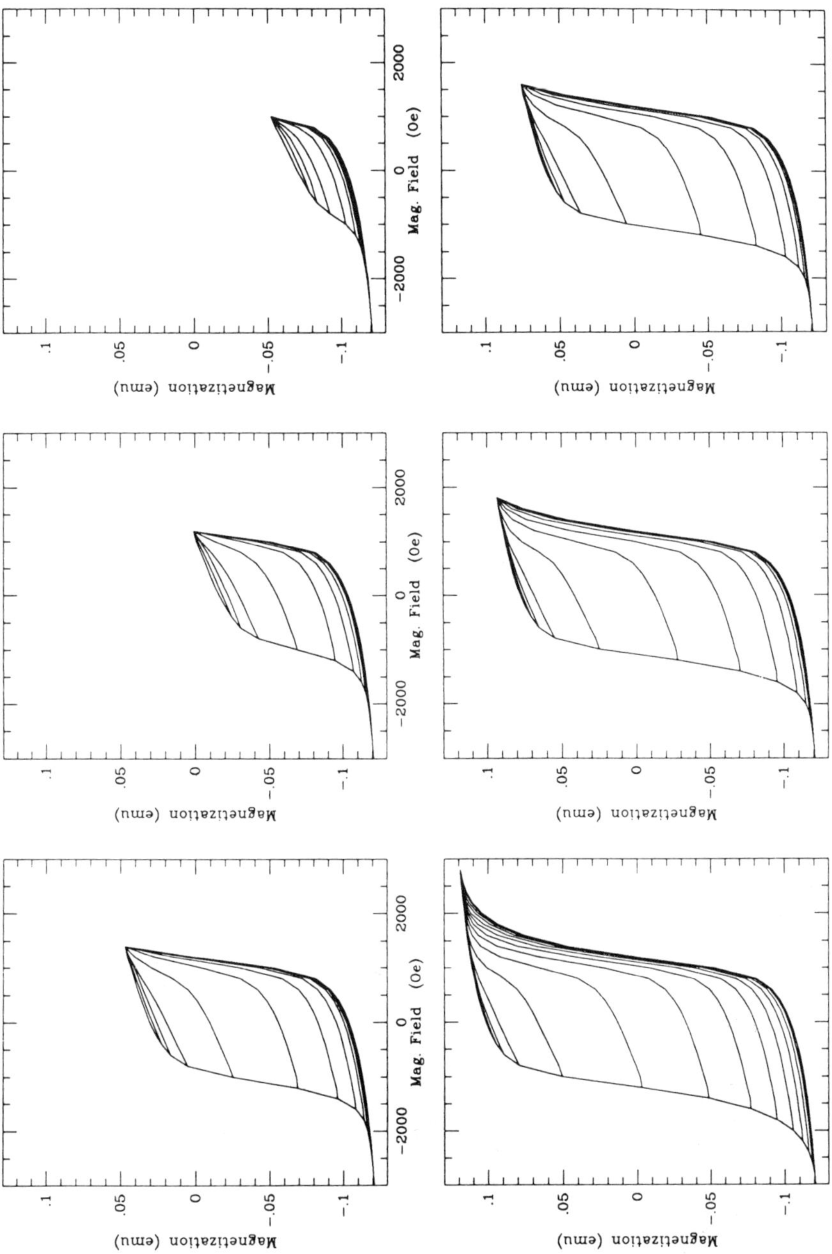

Fig. 2.61

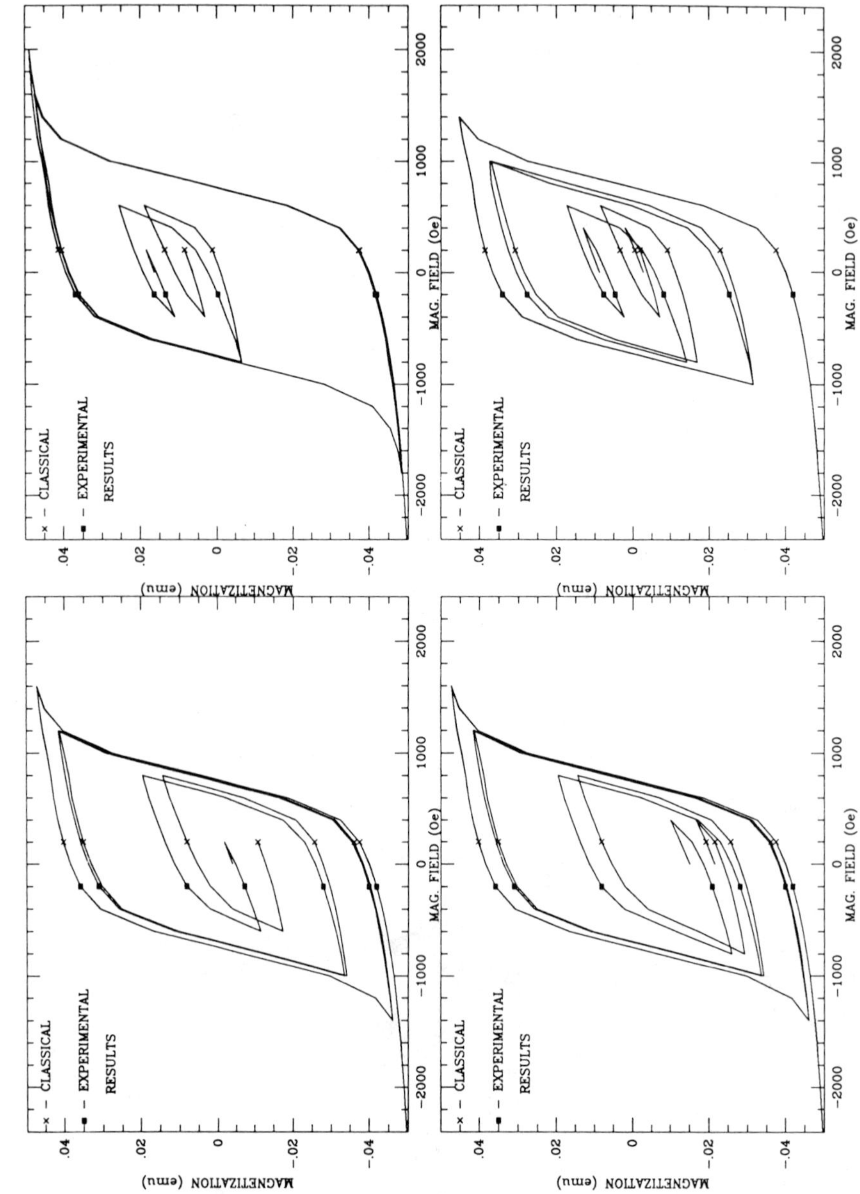

× – CLASSICAL
■ – EXPERIMENTAL
RESULTS
MAGNETIZATION (emu)
MAG. FIELD (Oe)
.04
.02
0
-.02
-.04
-2000
-1000
0
1000
2000

Fig. 2.62

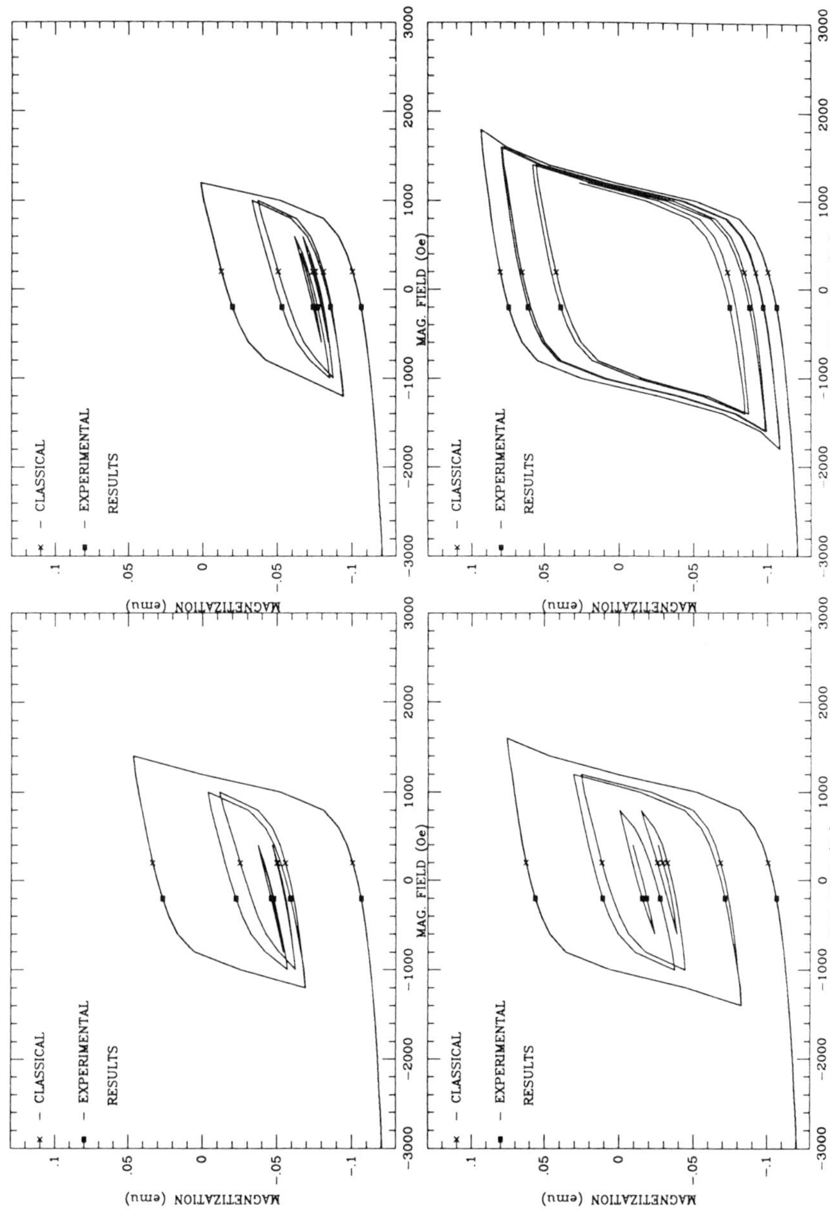

× – CLASSICAL
■ – EXPERIMENTAL
RESULTS
MAGNETIZATION (emu)
MAG. FIELD (Oe)
.1
.05
0
-.05
-.1
-3000
-2000
-1000
0
1000
2000
3000

Fig. 2.63

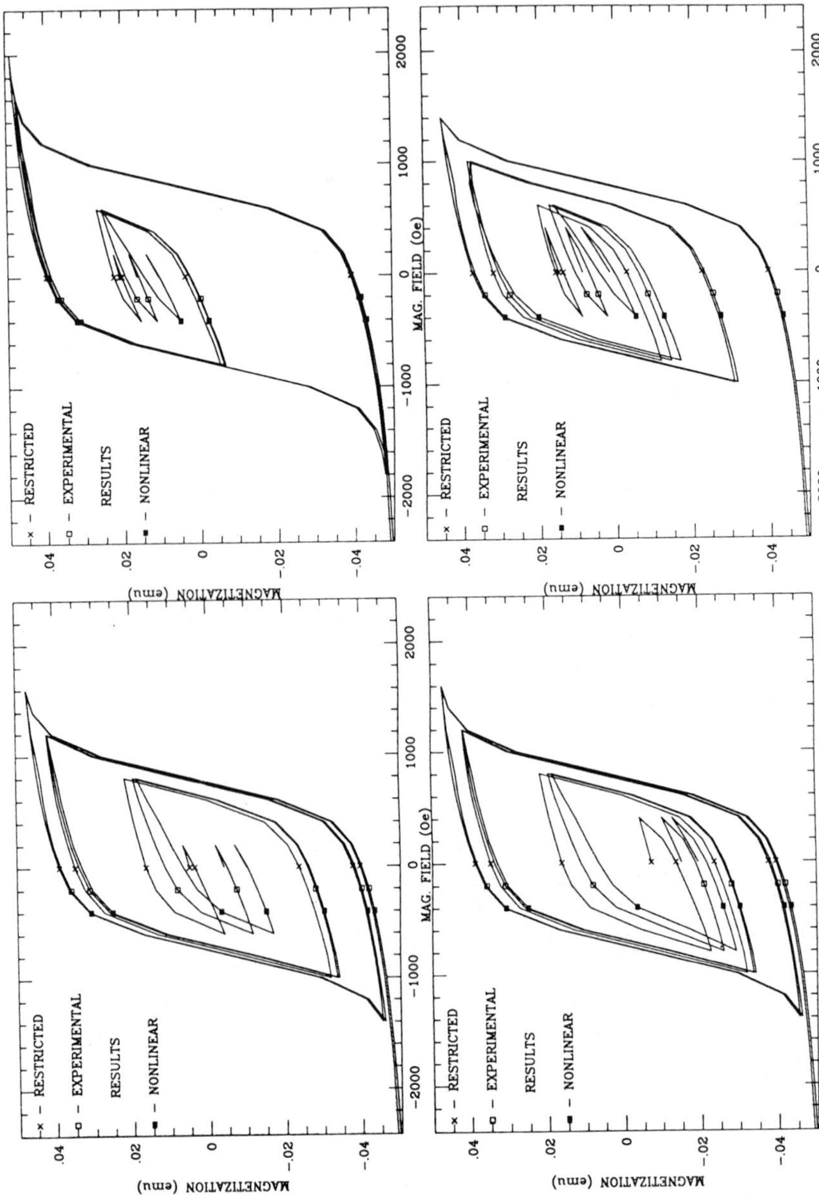
x – RESTRICTED
□ – EXPERIMENTAL
RESULTS
■ – NONLINEAR
MAGNETIZATION (emu)
MAG. FIELD (Oe)
.04
.02
0
-.02
-.04
-2000
-1000
0
1000
2000

Fig. 2.64

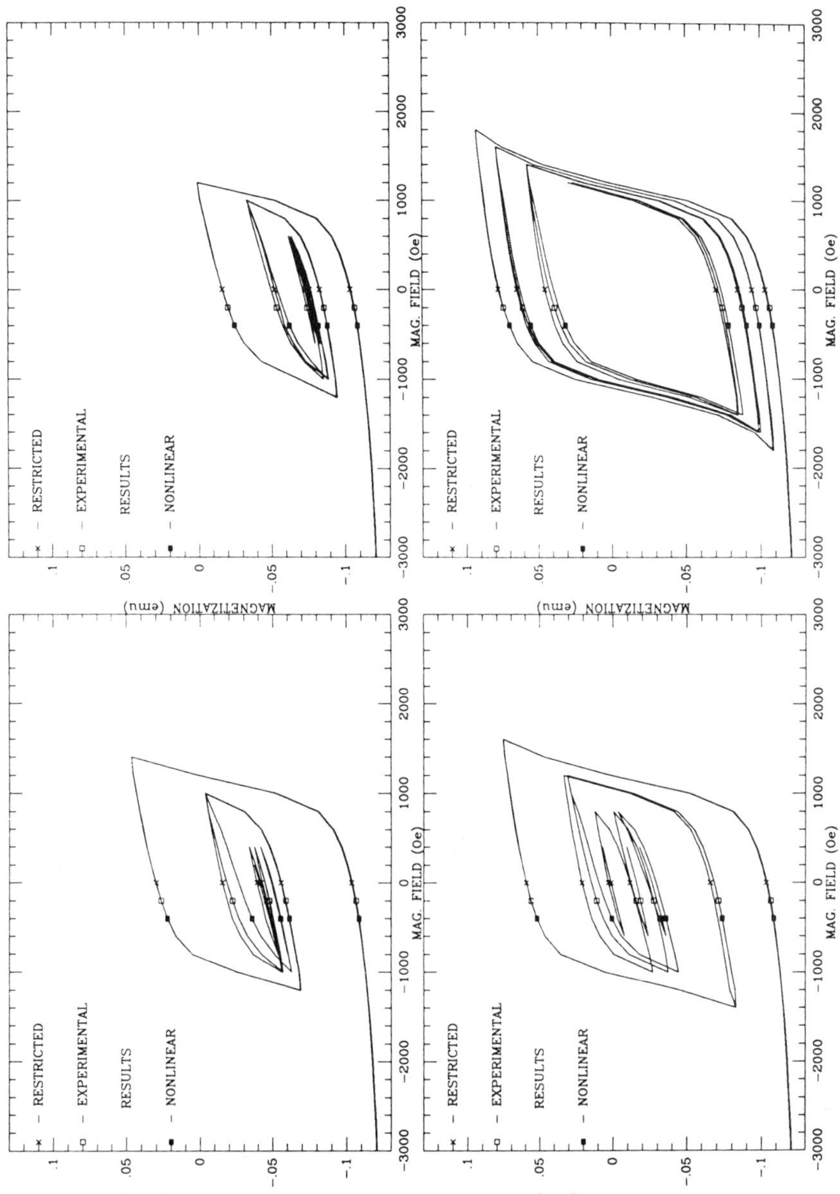
RESTRICTED
EXPERIMENTAL
RESULTS
NONLINEAR
MAGNETIZATION (emu)
MAG. FIELD (Oe)
-3000
-2000
-1000
0
1000
2000
3000
.1
.05
0
-.05
-.1

Fig. 2.65

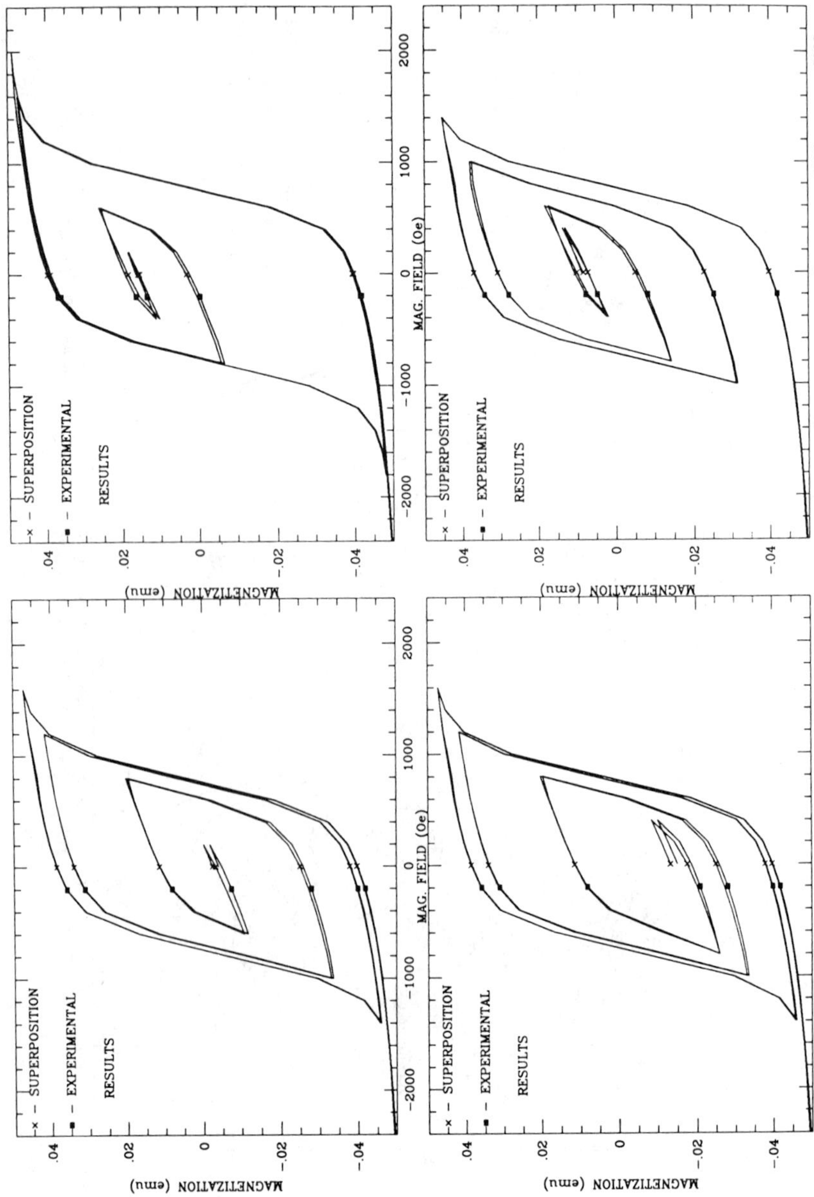
× – SUPERPOSITION
■ – EXPERIMENTAL
RESULTS
MAG. FIELD (Oe)
MAGNETIZATION (emu)

Fig. 2.66

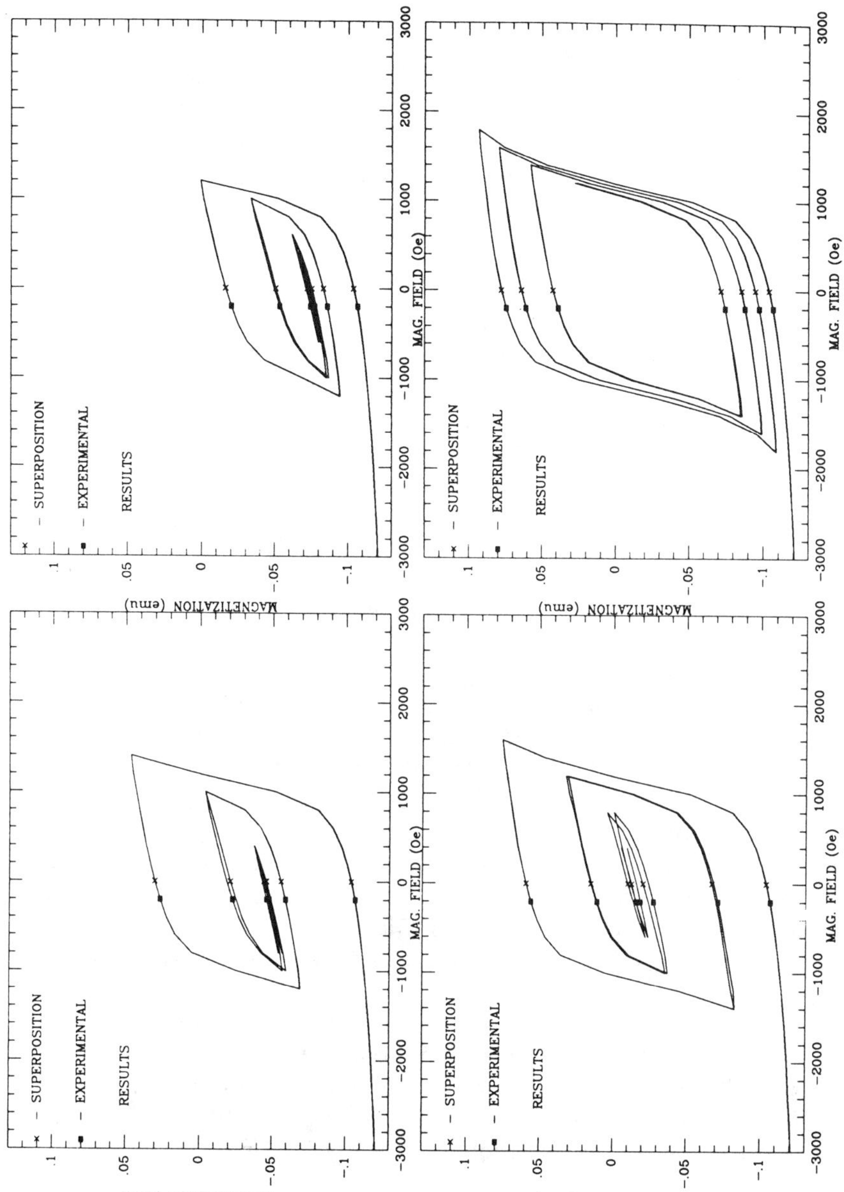
SUPERPOSITION
EXPERIMENTAL
RESULTS
MAGNETIZATION (emu)
MAG. FIELD (Oe)

Fig. 2.67

Preisach model (2.1) with the experimentally measured higher-order transition curves is demonstrated in Figs. 2.62 and 2.63. Similar comparisons for the nonlinear Preisach model (2.26) and the restricted Preisach model (2.86) are shown in Figs. 2.64 and 2.65. It is evident from these figures that the actual (experimentally measured) transition curves are "sandwiched" between the predictions of the nonlinear and restricted Preisach models. This observation has prompted the idea to use a "superposition" model

$$f(t) = \frac{1}{2}\iint [\mu(\alpha, \beta, u(t)) + \mu(\alpha, \beta, M_1)]\hat{\gamma}_{\alpha\beta}u(t)\,d\alpha\,d\beta + \frac{f^+_{u(t)} + f^-_{u(t)} + f_{M_1} + f_-}{4}. \qquad (2.292)$$

The numerical implementation of this model can be carried out by using the formula

$$f(t) = \frac{f_- + f^-_{u(t)}}{2} + \frac{1}{2}\sum_{k=1}^{n(t)} [P(M_{k+1}, m_k, u(t)) - P(M_k, m_k, u(t))] + \sum_{k=1}^{n(t)} [F(M_k, m_{k-1}, M_1) - F(M_k, m_k, M_1)], \qquad (2.293)$$

which follows from (2.66) and (2.113).

The "superposition" model turned out to be remarkably accurate for all used (and densely distributed) sequences of reversal values of magnetic fields. The comparison of the higher-order transition curves predicted by this model with the experimentally measured higher-order transition curves is demonstrated in Figs. 2.66 and 2.67.

The author wants to acknowledge that the experimental data used for the above comparisons were measured at the Center for Magnetic Recording Research, University of California at San Diego. This experimental work was carried out by my student G. Friedman and by C. Salling, a student of Professor S. Shultz.

It is worthwhile noting that the interest in the Preisach-type model is steadily growing. As a result, more and more attempts are being made to experimentally investigate the accuracy of this model. For instance, the accuracy of the classical Preisach model for the description of hysteresis of new magnetostrictive materials (Terfenol) has recently been investigated [54].

CHAPTER 3

Vector Preisach Models of Hysteresis

1 Classical Stoner-Wohlfarth Model of Vector Hysteresis

In magnetics, research on the modelling of scalar and vector hysteresis has been pursued along two quite distinct lines. Modeling of scalar hysteresis has been dominated by the Preisach approach. This approach can be traced back to a landmark paper [12]; it has proved to be very successful and has won many followers. On the other hand, phenomenological modelling of vector hysteresis has long been centered around the classical Stoner-Wohlfarth (S-W) model [3]. As a result, this model has been further developed and used in the area of magnetic recording [55–57]. The attractiveness of the S-W model can be attributed to its strong appeal to physical intuition. This appeal is, in turn, based on the fact that the S-W model is designed as an ensemble of single-domain, uniaxial magnetic particles. Since these particles have some features of physical realities, the S-W model is usually regarded as a physical (not mathematical) model. Due to its popularity in magnetics, the S-W model is a natural benchmark for comparison with other vector hysteresis models. This is the main reason why we precede our discussion of vector Preisach models by the discussion of the S-W model.

Since single-domain, uniaxial magnetic particles are the main building blocks of the S-W model, we begin with the discussion of hysteresis of these particles.

We consider a single-domain, uniaxial magnetic particle with magnetization (magnetic moment) $\vec{M}$ which may change its orientation under the influence of an applied field but has a constant magnitude. Such a particle is now commonly called a Stoner-Wohlfarth (S-W) particle. It is clear from the symmetry consideration that the vector $\vec{M}$ of this particle lies in the plane formed by the easy axis x and the applied magnetic field $\vec{H}$ (see Fig. 3.1). The orientation-dependent part of free energy $\mathscr{E}$ of the S-W particle is given by

$$\mathscr{E} = K \sin^2 \theta - \vec{M} \cdot \vec{H}, \tag{3.1}$$

where K is the anisotropy constant and θ is the angle between the easy axis and $\vec{M}$. The first term in the right-hand side of (3.1) represents the anisotropy

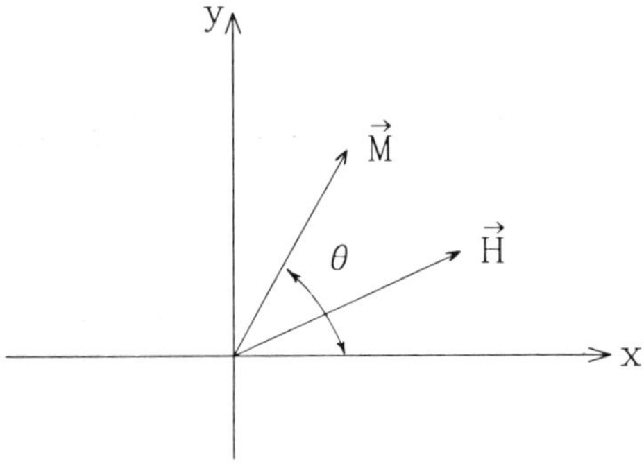

Fig. 3.1

energy, while the second term in (3.1) is the energy of interaction of magnetic moment $\vec{M}$ with the applied magnetic field.

By using the cartesian coordinates shown in Fig. 3.1, the expression (3.1) can be represented as

$$\mathscr{E} = K \sin^2 \theta - MH_x \cos \theta - MH_y \sin \theta. \tag{3.2}$$

Equilibrium orientations of $\vec{M}$ correspond to minima of $\mathscr{E}$, and they can be found from the equations

$$\frac{\partial \mathscr{E}}{\partial \theta} = 0, \tag{3.3}$$

$$\frac{\partial^2 \mathscr{E}}{\partial \theta^2} \geq 0. \tag{3.4}$$

From (3.2) and (3.3), we derive

$$2K \sin \theta \cos \theta + MH_x \sin \theta - MH_y \cos \theta = 0. \tag{3.5}$$

By introducing the so-called switching field α:

$$\alpha = \frac{2K}{M}, \tag{3.6}$$

the expression (3.5) can be rewritten as

$$\alpha \sin \theta \cos \theta + H_x \sin \theta - H_y \cos \theta = 0, \tag{3.7}$$

which is equivalent to

$$\frac{Hy}{\sin \theta} - \frac{Hx}{\cos \theta} = \alpha. \tag{3.8}$$

The equation (3.7) (as well as (3.8)) is a quartic equation with respect to $\cos \theta$. For this reason, this equation may have two or four real solutions. Which of these two cases is realized depends on the applied magnetic field, $\vec{H}$. In the first case, we have only one minimum and, consequently, only one equilibrium orientation of $\vec{M}$. In the second case, there are two minima and this results in two equilibrium orientations of $\vec{M}$. Thus, on $\vec{H}$-plane there are two different regions where $\vec{M}$ has one and two equilibrium orientations, respectively. On the boundary between the above two regions one minimum and one maximum merge together. In the case of minimum, the inequality (3.4) holds, while for a maximum we have $\partial^2\mathscr{E}/\partial\theta^2 \leq 0$. Consequently, on the above boundary we have

$$\frac{\partial \mathscr{E}}{\partial \theta} = 0 \quad \text{and} \quad \frac{\partial^2 \mathscr{E}}{\partial \theta^2} = 0. \tag{3.9}$$

The first condition in (3.9) leads to the equation (3.8). Now, let us derive the second equation by using both conditions in (3.9). From (3.2), by using the notation (3.6), we derive

$$\frac{1}{\sin\theta\cos\theta}\frac{\partial \mathscr{E}}{\partial \theta} = \alpha + \frac{H_x}{\cos\theta} - \frac{H_y}{\sin\theta}. \tag{3.10}$$

By differentiating (3.10), we obtain

$$\frac{\partial}{\partial\theta}\left(\frac{1}{\sin\theta\cos\theta}\right)\frac{\partial \mathscr{E}}{\partial\theta} + \frac{1}{\sin\theta\cos\theta}\frac{\partial^2\mathscr{E}}{\partial\theta^2} = -\frac{H_x\sin\theta}{\cos^2\theta} - \frac{H_y\cos\theta}{\sin^2\theta}. \tag{3.11}$$

By using both conditions (3.9), from (3.11) we find

$$\frac{H_x}{\cos^3\theta} + \frac{H_y}{\sin^3\theta} = 0. \tag{3.12}$$

Thus, on the boundary which separates regions with one and two minima, equations (3.8) and (3.12) are satisfied. These are two linear equations with respect to H_x and H_y. By solving these equations, we find

$$H_x = -\alpha\cos^3\theta, \qquad H_y = \alpha\sin^3\theta. \tag{3.13}$$

From (3.13) we find the equation for the boundary separating the above two regions:

$$H_x^{2/3} + H_y^{2/3} = \alpha^{2/3}. \tag{3.14}$$

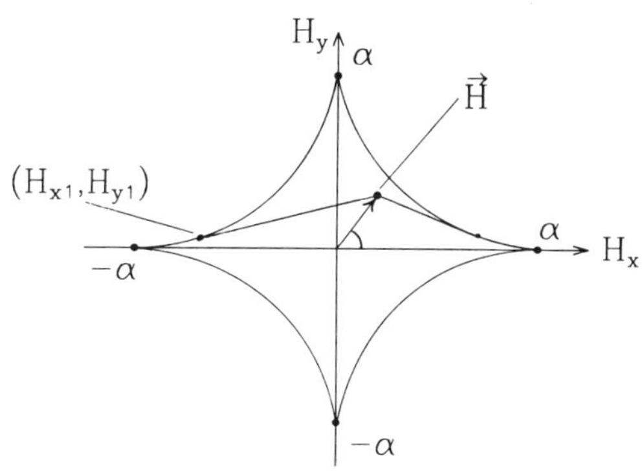

Fig. 3.2

This equation represents the astroid curve shown in Fig. 3.2. This astroid curve helps to visualize the solution of the quartic equation (3.7). This solution can be found by using the following geometric construction [58], [59]. For a given external magnetic field $\vec{H}$ with components H_x and H_y the directions of $\vec{M}$ which satisfy the equation (3.7) are parallel to those tangent lines to the astroid which pass through the point $\vec{H}$ (see Fig. 3.2). The proof of the above statement proceeds as follows.

Let H_x and H_y be cartesian components of $\vec{H}$, and let H_{x_1} and H_{y_1} be cartesian coordinates of the point on the astroid at which the above-mentioned tangent line touches the astroid. For the slope of this tangent line we have

$$m = \frac{H_{y_1} - H_y}{H_{x_1} - H_x}, \tag{3.15}$$

which is tantamount to

$$H_{y_1} - H_y = m(H_{x_1} - H_x). \tag{3.16}$$

Since the point (H_{x_1}, H_{y_1}) belongs to the astroid, we have

$$H_{x_1}^{2/3} + H_{y_1}^{2/3} = \alpha^{2/3}. \tag{3.17}$$

By using implicit differentiation of (3.17) with respect to H_{x_1}, we derive

$$\frac{2}{3} H_{x_1}^{-1/3} + \frac{2}{3} H_{y_1}^{-1/3} \frac{dH_{y_1}}{dH_{x_1}} = 0. \tag{3.18}$$

From (3.18), we obtain

$$m = \frac{dH_{y_1}}{dH_{x_1}} = -\left(\frac{H_{y_1}}{H_{x_1}}\right)^{1/3}. \tag{3.19}$$

Let β be the angle formed by the above tangent line with the easy axis of the particle. Then,

$$m = \tan \beta, \tag{3.20}$$

and from (3.19) and (3.20) we find

$$\frac{H_{y_1}}{H_{x_1}} = -\tan^3 \beta. \tag{3.21}$$

Expressions (3.17) and (3.21) can be construed as two simultaneous equations with respect to two unknowns: H_{x_1} and H_{y_1}. By solving these equations, we obtain

$$H_{x_1} = -\alpha \cos^3 \beta, \qquad H_{y_1} = \alpha \sin^3 \beta. \tag{3.22}$$

By substituting (3.20) and (3.22) into (3.16), we find

$$\alpha \sin^3 \beta - H_y = \tan \beta(-\alpha \cos^3 \beta - H_x). \tag{3.23}$$

A trivial transformation leads to

$$\alpha(\sin^3 \beta \cos \beta + \cos^3 \beta \sin \beta) + H_x \sin \beta - H_y \cos \beta = 0, \tag{3.24}$$

which can bc simplified as

$$\alpha \sin \beta \cos \beta + H_x \sin \beta - H_y \cos \beta = 0. \tag{3.25}$$

It is clear that the equations (3.7) and (3.25) are identical. Consequently,

$$\beta = \theta. \tag{3.26}$$

This proves the validity of the above-described geometric construction.

It can be shown that equilibrium orientations of $\vec{M}$ correspond to the tangent lines with smallest slopes. It is clear that, when the point $\vec{H}$ is outside the astroid, only two tangent lines are possible, and therefore there is only one equilibrium orientation of $\vec{M}$. When $\vec{H}$ is inside the astroid, there are four tangent lines. However, only two of these tangent lines represent equilibrium orientations of the magnetization $\vec{M}$. Which one is realized depends on the previous history of the magnetization.

The described geometric rules allow one to compute hysteresis loops of a S-W particle for the case when the applied magnetic field is restricted to vary along one arbitrarily chosen direction. Suppose that this direction is specified by the line $a - a'$ (see Fig. 3.3) and that the magnetic field is first monotonically increased from its value H_- corresponding to the point 1 to the value H_+ corresponding to the point 6 and then is monotonically decreased back to H_-. The dependence of the projection of magnetization on the line $a - a'$ on the value of the magnetic field H exhibits hysteresis which is shown in Fig. 3.4. This is clear from the Fig. 3.3. Indeed, as we move up along the line $a - a'$, the equilibrium (stable) orientations of $\vec{M}$ coincide with the directions of tangent lines to the right-hand side of the astroid until we reach the point 5. At this point a "switch" from the right-hand side to the left-hand side of the astroid occurs, and after that point stable orientations of $\vec{M}$ coincide with directions of tangent lines to the latter part of the astroid. As we move down along the line $a - a'$ from the point 6, we use the tangent lines to the left-hand side of the astroid to determine the stable direction of $\vec{M}$. However, at point 2 a "switch" from the left-hand side to the right-hand side of the astroid occurs

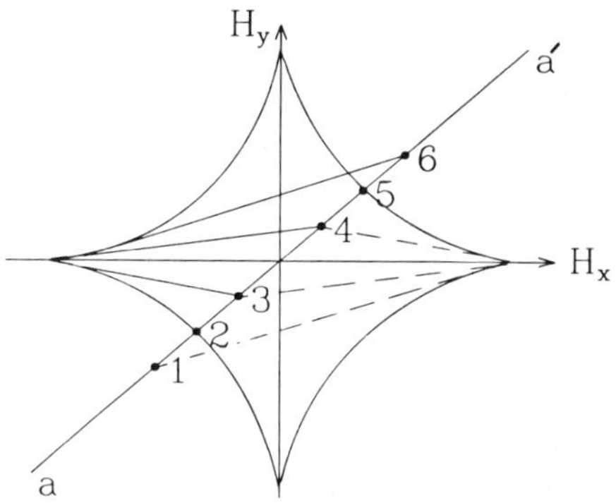

Fig. 3.3

and after that point stable orientations of $\vec{M}$ coincide with directions of tangent lines to the latter part of the astroid. Thus, for the points of the line $a - a'$ which are inside the astroid, there are two stable orientations of $\vec{M}$ which result in two different branches of the hysteresis loop shown in Fig. 3.4. It is clear from the preceding discussion that if the applied field varies along the easy axis x than a S-W particle exhibits a rectangular hysteresis loop as shown in Fig. 3.5. It is also clear that if the applied magnetic field is varied along the direction perpendicular to the easy axis, then due to the symmetry there is no hysteresis effect and a S-W particle exhibits a single valued magnetization curve shown in Fig. 3.6. Thus, the shape of the hysteresis loops depends on the direction along which the applied field is being varied.

It has been shown in Chapter II (see formula (2.81)) that the loop shown in Fig. 3.4 can be represented in terms of the rectangular $\hat{\gamma}$-loop. Recently, my student G. Freedman found a very interesting generalization of the formula (2.81) (see [61]). This generalization represents the magnetization of S-W particles for arbitrary (not only collinear) variations of the magnetic field in terms of rectangular $\hat{\gamma}$-loops. The basis for this representation is the notion that there are two distinct states (vector branches) for any S-W particle. In the first state, stable orientations of the magnetization coincide with directions

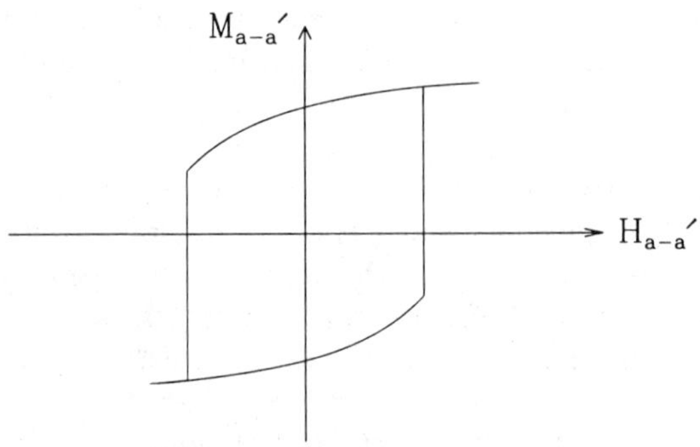

Fig. 3.4

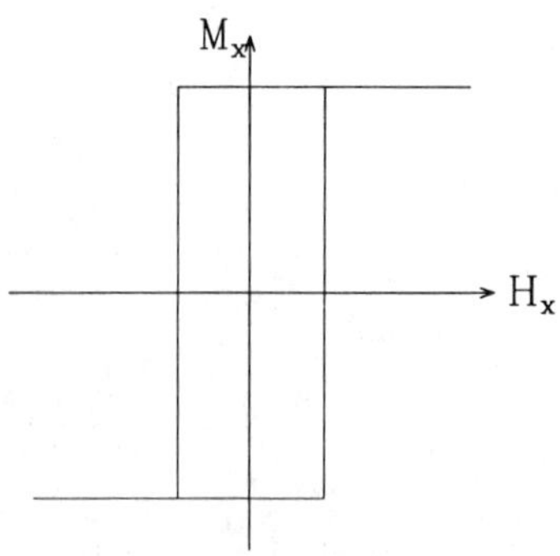

Fig. 3.5

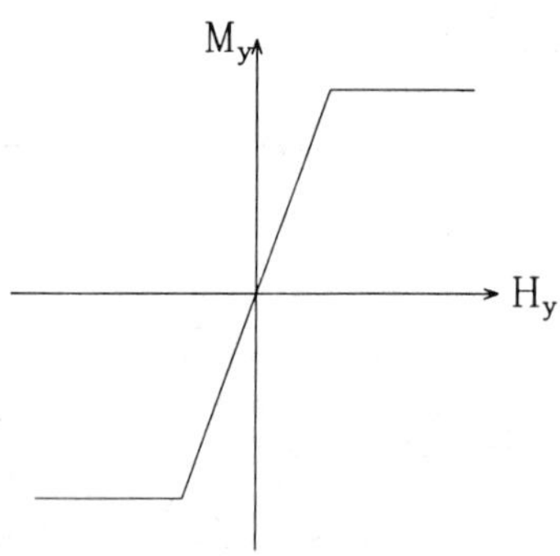

Fig. 3.6

of tangent lines to the left-hand side of the astroid. The notation $\vec{M}^+(\varphi(t), H(t))$ will be used for the magnetization in the first state, where $\varphi(t)$ is the angle formed by the applied magnetic field with the easy axis and $H(t)$ is the magnitude of magnetic field. In the second state, the stable orientations of $\vec{M}$ coincide with directions of tangent lines to the right-hand side of the astroid. The notation $\vec{M}^-(\varphi(t), H(t))$ will be used for the magnetization in the second state. It is clear that $\vec{M}^+(\varphi(t), H(t))$ and $\vec{M}^-(\varphi(t), H(t))$ can be found geometrically by using the previously described "astroid" rule or by solving the quartic equation (3.7). The rectangular loop operator $\hat{\gamma}_{\alpha,-\alpha}$ will be employed to describe switching from the first state to the second state and vice versa. The input $v(t)$ for this operator is given by

$$v(t) = \frac{\cos\varphi(t)}{|\cos\varphi(t)|}[H_x(t)^{2/3} + H_y(t)^{2/3}]^{3/2}. \tag{3.27}$$

It is clear from (3.27) and (3.14) that $v(t)$ reaches the value α as the tip of $\vec{H}(t)$ crosses the right-hand side of the astroid, and $v(t)$ reaches the value $-\alpha$ as the tip of $\vec{H}(t)$ crosses the left-hand half of the astroid. This shows that the switching of the rectangular loop $\hat{\gamma}_{\alpha,-\alpha}$ occurs at the same time as the switching of the S-W particle from one state to another. By using this fact, we can represent the magnetization $\vec{M}(t)$ of the S-W particle as

$$\vec{M}(t) = \vec{c}(t)\hat{\gamma}_{\alpha,-\alpha}v(t) + \vec{d}(t), \tag{3.28}$$

where

$$\vec{c}(t) = \tfrac{1}{2}[\vec{M}^+(\varphi(t), H(t)) - \vec{M}^-(\varphi(t), H(t))], \tag{3.29}$$

$$\vec{d}(t) = \tfrac{1}{2}[\vec{M}^+(\varphi(t), H(t)) + \vec{M}^-(\varphi(t), H(t))]. \tag{3.30}$$

It is clear from (3.28), (3.29) and (3.30) that $\vec{M}(t) = \vec{M}^+(\varphi(t), H(t))$ when $\hat{\gamma}_{\alpha,-\alpha}v(t) = 1$, and $\vec{M}(t) = \vec{M}^-(\varphi(t), H(t))$ when $\hat{\gamma}_{\alpha,-\alpha}v(t) = -1$. Switchings of $\hat{\gamma}_{\alpha,-\alpha}v(t)$ from 1 to -1 and vice versa occur at the times when the tip of $\vec{H}(t)$ crosses the astroid. This proves that formulas (3.27)–(3.30) give the right representation for the magnetization of the S-W particle.

Having described the basic properties of a S-W particle, we can now proceed to the discussion of the S-W hysteresis model. This model is designed as an ensemble of S-W particles. Consider an infinite set of S-W particles with different orientations of their easy axis and different values of switching field α. The notation $\hat{S}_{\theta,\alpha}$ will be used for a S-W particle whose switching field is equal to α and whose easy axis forms the angle θ with the x-axis. By using this notation, the S-W model can be represented mathematically as

$$\vec{M}(t) = \iint \xi(\theta, \alpha)\hat{S}_{\theta,\alpha}\vec{H}(t)d\theta d\alpha, \tag{3.31}$$

where $\xi(\theta, \alpha)$ is a distribution function which should be determined by fitting the model to some experimental data.

The expression (3.31) defines the S-W model in terms of magnetic quantities such as magnetization $\vec{M}$ and magnetic field $\vec{H}(t)$. However, it is possible to interpret the S-W model as a general mathematical model of vector hysteresis by writing this model in the form

$$\vec{f}(t) = \iint \xi(\theta, \alpha)\hat{S}_{\theta,\alpha}\vec{u}(t)d\theta d\alpha, \tag{3.32}$$

where $\vec{f}(t)$ is the vector output, while $\vec{u}(t)$ is the vector input.

The output of the operator $\hat{S}_{\theta,\alpha}$ can be formally determined by using the astroid rule or the quartic equation (3.7) in which H_x and H_y are replaced by $u_x(t)$ and $u_y(t)$, respectively. In other words, the S-W model can be defined in purely mathematical terms without using any connections of this model to some physical objects such as uniaxial, single-domain magnetic particles. Such a purely mathematical point of view of the S-W model may have a twofold advantage. First, it suggests some possibilities of using this model in areas other than magnetics. Second, it may open some opportunities for further generalization of this model.

By using representation (3.28)–(3.30) for the S-W particles, the S-W model (3.31) (or (3.32)) can be written in terms of $\hat{\gamma}$-operators. This suggests some connections between the S-W model and the Preisach-type models. In particular, Preisach-type diagrams can be used to keep track of the switching of different S-W particles.

The S-W model has been known and used in magnetics for a long time. Gradually, it has been realized that it has some serious deficiencies. The most important of them can be summarized as follows.

Since the S-W model is designed as an ensemble (superposition) of particles (hysteresis nonlinearities) with symmetric loops, this model does not describe nonsymmetric minor loops. This deficiency is often attributed to the fact that the S-W model does not account for "particle interactions." This deficiency can be somewhat corrected by expanding the set of elementary hysteresis operators $\hat{S}_{\theta,\alpha}$ and by introducing the operators $\hat{S}_{\theta,\alpha,\vec{a}}$ with shifted astroids (see Fig. 3.7). Then, the generalized S-W model can be represented as

$$\vec{f}(t) = \iiiint \xi(\theta, \alpha, \vec{a})\hat{S}_{\theta,\alpha,\vec{a}}u(t)d\theta d\alpha d\vec{a}. \tag{3.33}$$

However, this generalization will require much more computational work for the numerical implementation of the S-W model. Even without this general-

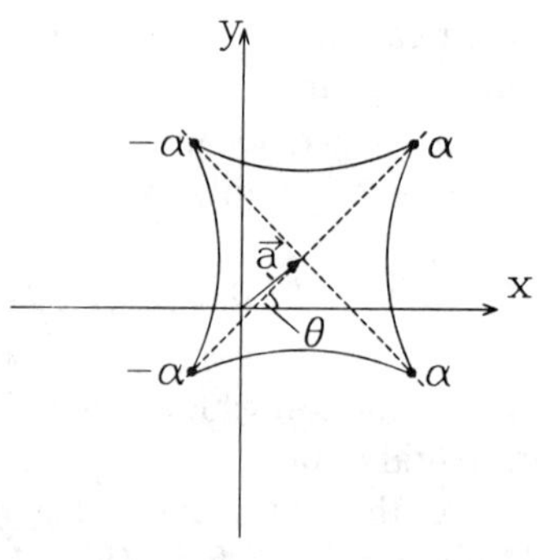

Fig. 3.7

ization, the S-W model is computationally slow. This is in part because the calculations of outputs of individual elementary operators $\hat{S}_{\theta,\alpha}$ require the solution of quartic equations associated with the astroid construction. In addition, individual outputs should be integrated over some distributions of S-W nonlinearities (particles) with respect to their easy axis directions and switching fields. This requires the evaluation of double integrals in the case of the classical 2-D S-W model (3.32). The last difficulty is magnified in the case of the generalized S-W model (3.33).

Furthermore, the identification problem of finding the distribution function $\xi(\theta, \alpha)$ by fitting the S-W model to some experimental data has not been adequately addressed yet. Solutions to this problem are usually achieved by some artwork rather than by using a well-established procedure.

Finally, the design of the S-W model does not address the question of what part of the past input history leaves its mark upon future output values of vector hysteresis nonlinearities. This question is very important for the understanding of the phenomenological nature of vector hysteresis.

Our research has been motivated by the desire to circumvent the deficiencies of the S-W model described above. To achieve this goal, we have turned to the Preisach approach and tried to extend it to the vector case. The guiding idea in our efforts has been the notion that the scalar hysteresis is a particular case of vector hysteresis. As a result, many important and characteristic properties of vector hysteresis can be exhibited in the scalar case. By exploring this notion, new vector Preisach models of hysteresis have been developed. These models have many of the desirable features of the scalar Preisach hysteresis models, and they represent a viable alternative to the S-W model.

2 Definition of Vector Preisach Models of Hysteresis and Their Numerical Implementation

For the sake of generality, a vector hysteresis nonlinearity will be characterized below by a vector input $\vec{u}(t)$ and a vector output $\vec{f}(t)$. In magnetic applications $\vec{u}(t)$ is the magnetic field, while $\vec{f}(t)$ is the magnetization. The most immediate problem we face is how to define vector hysteresis in a mathematically rigorous as well as physically meaningful way. It is not clear if these two requirements can currently be reconciled, but at least we shall try to achieve some clarity in this matter. First, we need to understand what part of the past input history affects the future values of the output. The starting point for our reasoning is the fact that scalar hysteresis is a particular case of vector hysteresis. This case is realized when the vector input $\vec{u}(t)$ is restricted to vary along only one direction. In fact, it can even be argued successfully (at least in the area of magnetics) that there is no such phenomenon as scalar hysteresis. Whenever we talk about scalar hysteresis, we actually deal with some specific properties of vector hysteresis which have been observed for input variations along some fixed directions. Thus, it can be concluded that vector hysteresis is reduced to scalar hysteresis when the vector input is restricted to vary along any arbitrarily fixed direction. Since the above reduction occurs for *any* direction and since past extremum values of input constitute the essential part of past history for scalar hysteresis, it is natural to assume that in the case of vector hysteresis past extremum values of input projections along all possible directions may leave their mark upon the future. As a result, we arrive at the following definition of vector hysteresis. Vector hysteresis is a vector nonlinearity with the property that past extremum values of input projections along all possible directions may affect future values of the output. Since no mention is made of the speed of input variations, it is apparent that the above definition implies static vector hysteresis. According to the given definition of vector hysteresis, mathematical models of vector hysteresis should be able to detect and store past extrema of input projections along all possible directions and choose the appropriate value of the vector output according to the accumulated history. To detect and accumulate the past extremum values of input projections along all possible directions, the scalar Preisach models (Preisach's particles) can be employed. These scalar models are continuously distributed along all possible directions (see Fig. 3.8). Thus, scalar Preisach models are main building blocks for the vector model, which is constructed as a superposition of scalar models. This can be expressed mathematically in two dimensions as

$$\vec{f}(t) = \oint_{|\vec{r}|=1} \vec{r}\,\hat{\Gamma}_{\vec{r}}(\vec{r}\cdot\vec{u}(t))\,d\ell_r, \tag{3.34}$$

and the integration in (3.34) is performed over a unit circle.

Similarly, a 3-D vector Preisach model can be written in the form

$$\vec{f}(t) = \oiint_{|\vec{r}|=1} \vec{r}\,\hat{\Gamma}_{\vec{r}}(\vec{r}\cdot\vec{u}(t))\,ds_r, \tag{3.35}$$

where the integration is performed over a unit sphere.

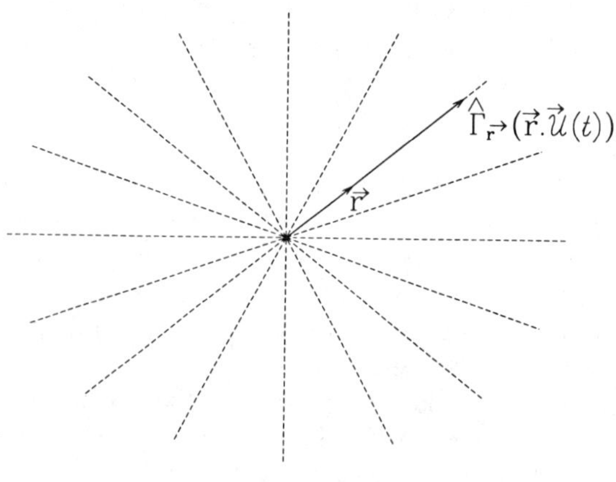

Fig. 3.8

The scalar Preisach models $\hat{\Gamma}_{\vec{r}}$ are defined by

$$\hat{\Gamma}_{\vec{r}}(\vec{r}\cdot\vec{u}(t)) = \iint_{\alpha\geq\beta} \nu(\alpha, \beta)\hat{\gamma}_{\alpha\beta}(\vec{r}\cdot\vec{u}(t))d\alpha d\beta \tag{3.36}$$

for isotropic vector models, and

$$\hat{\Gamma}_{\vec{r}}(\vec{r}\cdot\vec{u}(t)) = \iint_{\alpha\geq\beta} \nu(\alpha, \beta, \vec{r})\hat{\gamma}_{\alpha\beta}(\vec{r}\cdot\vec{u}(t))d\alpha d\beta \tag{3.37}$$

for anisotropic vector models.

Some ideas concerning the construction of vector Preisach models which are somewhat similar to the concept described above have been briefly mentioned (without any analytical details) in [2], [21]. Some similarities can also be found between our definition of the vector Preisach models and a purely mathematical vector generalization of the scalar Preisach model discussed in [22].

The following theorem further elucidates the above definition of vector Preisach models.

Theorem

The integration in (3.35) *over a unit sphere can be reduced to the integration over a unit hemisphere.*

Proof

We partition the unit sphere into two hemisphere C^+ and C^-. For every point $\vec{r}^{\,+} \in C^+$ there is a corresponding opposite point $\vec{r}^{\,-} \in C^-$ (see Fig. 3.9)

$$\vec{r}^{\,+} = -\vec{r}^{\,-}. \tag{3.38}$$

From (3.35) we find

$$\vec{f}(t) = \iint_{C^+} \vec{r}^{\,+}\hat{\Gamma}_{\vec{r}^{\,+}}(\vec{r}^{\,+}\cdot\vec{u}(t))ds_+ + \iint_{C^-} \vec{r}^{\,-}\hat{\Gamma}_{\vec{r}^{\,-}}(\vec{r}^{\,-}\cdot\vec{u}(t))ds_-, \tag{3.39}$$

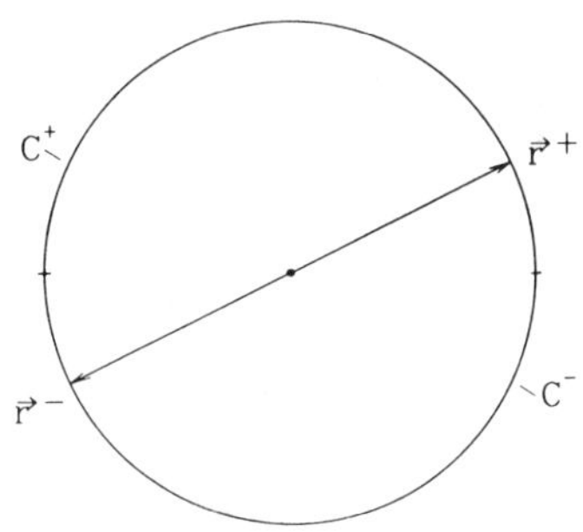

Fig. 3.9

where

$$\hat{\Gamma}_{\vec{r}^+}(\vec{r}^+\cdot\vec{u}(t)) = \iint_{\alpha\geq\beta} \nu(\alpha,\beta,\vec{r}^+)\hat{\gamma}_{\alpha\beta}(\vec{r}^+\cdot\vec{u}(t))d\alpha d\beta, \tag{3.40}$$

$$\hat{\Gamma}_{\vec{r}^-}(\vec{r}^-\cdot\vec{u}(t)) = \iint_{\alpha\geq\beta} \nu(\alpha,\beta,\vec{r}^-)\hat{\gamma}_{\alpha\beta}(\vec{r}^-\cdot\vec{u}(t))d\alpha d\beta. \tag{3.41}$$

The following identity can be verified for any continuous piecewise monotonic function $v(t)$:

$$\hat{\gamma}_{\alpha\beta}v(t) = -\hat{\gamma}_{-\beta,-\alpha}(-v(t)). \tag{3.42}$$

Indeed, if $v(t) < \beta$, then $-v(t) > -\beta$ and

$$\hat{\gamma}_{\alpha\beta}v(t) = -1, \quad \text{while} \quad \hat{\gamma}_{-\beta,-\alpha}(-v(t)) = +1. \tag{3.43}$$

If $v(t)$ increases and exceeds α, then $-v(t)$ decreases and becomes smaller than $-\alpha$. Consequently,

$$\hat{\gamma}_{\alpha\beta}v(t) = 1, \quad \text{while} \quad \hat{\gamma}_{-\beta,-\alpha}(-v(t)) = -1. \tag{3.44}$$

Similarly, it can be shown that the identity (3.42) holds for any monotonic decrease of $v(t)$.

In the last integral in (3.39) we change the variable $\vec{r}^-$ to $-\vec{r}^+$. As a result, we find

$$\iint_{C^-} \vec{r}^-\hat{\Gamma}_{\vec{r}^-}(\vec{r}^-\cdot\vec{u}(t))ds_- = -\iint_{C^+} \vec{r}^+\hat{\Gamma}_{-\vec{r}^+}(-\vec{r}^+\cdot\vec{u}(t))ds_+, \tag{3.45}$$

where

$$\hat{\Gamma}_{-\vec{r}^+}(-\vec{r}^+\cdot\vec{u}(t)) = \iint_{\alpha\geq\beta} \nu(\alpha,\beta,-\vec{r}^+)\hat{\gamma}_{\alpha\beta}(-\vec{r}^+\cdot\vec{u}(t))d\alpha d\beta. \tag{3.46}$$

By using the identity (3.42), the last integral in (3.46) can be transformed as follows:

$$\iint_{\alpha\geq\beta} \nu(\alpha,\beta,-\vec{r}^+)\hat{\gamma}_{\alpha\beta}(-\vec{r}^+\cdot\vec{u}(t))d\alpha d\beta$$

$$= -\iint_{\alpha\geq\beta} \nu(\alpha,\beta,-\vec{r}^+)\hat{\gamma}_{-\beta,-\alpha}(\vec{r}^+\cdot\vec{u}(t))d\alpha d\beta. \tag{3.47}$$

In the last integral we will change α to $-\beta$ and β to $-\alpha$, then according to (3.46) and (3.47) we obtain

$$\hat{\Gamma}_{-\vec{r}^+}(-\vec{r}^+\cdot\vec{u}(t)) = -\iint_{\alpha\geq\beta} \nu(-\beta,-\alpha,-\vec{r}^+)\hat{\gamma}_{\alpha\beta}(\vec{r}^+\cdot\vec{u}(t))d\alpha d\beta. \tag{3.48}$$

By substituting (3.48) into (3.45), we find

$$\iint_{C^-} \vec{r}^{\,-} \hat{\gamma}_{\vec{r}^{\,-}}(\vec{r}^{\,-} \cdot \vec{u}(t)) ds_- = \iint_{C^+} \vec{r}^{\,+} \left(\iint_{\alpha \geq \beta} \nu(-\beta, -\alpha, -\vec{r}^{\,+}) \hat{\gamma}_{\alpha\beta}(\vec{r}^{\,+} \cdot \vec{u}(t)) d\alpha d\beta \right) ds_+. \quad (3.49)$$

By substituting (3.40) into the first integral in (3.39), we have

$$\iint_{C^+} \vec{r}^{\,+} \hat{\Gamma}_{\vec{r}^{\,+}}(\vec{r}^{\,+} \cdot \vec{u}(t)) ds_+ = \iint_{C^+} \vec{r}^{\,+} \left(\iint_{\alpha \geq \beta} \nu(\alpha, \beta, \vec{r}^{\,+}) \hat{\gamma}_{\alpha\beta}(\vec{r}^{\,+} \cdot \vec{u}(t)) d\alpha d\beta \right) ds_+. \quad (3.50)$$

From (3.39), (3.49), and (3.50), we obtain

$$\vec{f}(t) = \iint_{C^+} \vec{r}^{\,+} \left[\iint_{\alpha \geq \beta} (\nu(\alpha, \beta, \vec{r}^{\,+}) + \nu(-\beta, -\alpha, -\vec{r}^{\,+})) \hat{\gamma}_{\alpha\beta}(\vec{r}^{\,+} \cdot \vec{u}(t)) d\alpha d\beta \right] ds_+. \quad (3.51)$$

By introducing a new function:

$$\nu'(\alpha, \beta, \vec{r}^{\,+}) = \nu(\alpha, \beta, \vec{r}^{\,+}) + \nu(-\beta, -\alpha, -\vec{r}^{\,+}), \quad (3.52)$$

from (3.51) we derive

$$\vec{f}(t) = \iint_{C^+} \vec{r}^{\,+} \hat{\Gamma}'_{\vec{r}^{\,+}}(\vec{r}^{\,+} \cdot \vec{u}(t)) ds_+, \quad (3.53)$$

where

$$\hat{\Gamma}'_{\vec{r}^{\,+}}(\vec{r}^{\,+} \cdot \vec{u}(t)) = \iint_{\alpha \geq \beta} \nu'(\alpha, \beta, \vec{r}^{\,+}) \hat{\gamma}_{\alpha\beta}(\vec{r}^{\,+} \cdot \vec{u}(t)) d\alpha d\beta. \quad (3.54)$$

This completes the proof of the theorem.

The proven theorem suggests that (3.53) and (3.54) can be regarded as an equivalent definition of the vector Preisach model. This definition will be used in our subsequent discussions and the superscript $'$ will be omitted. It is also clear that a similar theorem is valid for the two-dimensional vector Preisach model (3.34) as well. Consequently, this model can be represented as

$$\vec{f}(t) = \int_{L_+} \vec{r}^{\,+} \hat{\Gamma}'_{\vec{r}^{\,+}}(\vec{r}^{\,+} \cdot \vec{u}(t)) d\ell_+, \quad (3.55)$$

where L_+ is a semicircle and $\hat{\Gamma}'_{\vec{r}^{\,+}}$ is defined by (3.54).

It is apparent from the proof of the theorem that particular choices of semispheres and semicircles in (3.53) and (3.55), respectively, are unimportant; all these choices will lead to equivalent vector Preisach models.

From the proven theorem we can also derive some interesting symmetry properties of the function ν'. Indeed, if in (3.52) we change α to $-\beta$, β to $-\alpha$, and $\vec{r}^+$ to $-\vec{r}^+$, then we find

$$\nu'(-\beta, -\alpha, -\vec{r}^+) = \nu(-\beta, -\alpha, -\vec{r}^+) + \nu(\alpha, \beta, \vec{r}^+). \tag{3.56}$$

From (3.56) and (3.52), we obtain

$$\nu'(\alpha, \beta, \vec{r}^+) = \nu'(-\beta, -\alpha, -\vec{r}^+). \tag{3.57}$$

This expresses the property of mirror symmetry of the function ν' with respect to the line $\alpha = -\beta$. This also extends the definition of the function ν' from a unit semisphere or unit semicircle to the entire unit sphere or unit circle, respectively.

In the case of isotropic vector models, the expression (3.57) can be simplified as

$$\nu'(\alpha, \beta) = \nu'(-\beta, -\alpha), \tag{3.58}$$

which is similar to the symmetry property of the μ-function for the classical scalar Preisach model (see the expression (1.46)).

Up to this point, the vector Preisach models have been defined in coordinate invariant forms (3.34), (3.35), (3.53), and (3.55). However, in many applications it is more convenient to use the expressions for the vector Preisach models in spherical and polar coordinates for three and two dimensions, respectively. In the case of spherical coordinates, we have

$$ds_+ = \sin\theta d\theta d\varphi, \quad \vec{r}^+ = \vec{e}_{\theta,\varphi}, \quad \vec{r}^+ \cdot \vec{u}(t) = u_{\theta,\varphi}(t), \tag{3.59}$$

where $\vec{e}_{\theta,\varphi}$ is a unit vector along the direction specified by angles φ and θ, and $u_{\theta,\varphi}(t)$ is the projection of $\vec{u}(t)$ along the direction of $\vec{e}_{\theta,\varphi}$.

By using (3.59), the 3-D vector Preisach model (3.53) and (3.54) can be written as

$$\vec{f}(t) = \int_0^{2\pi} \int_0^{\pi/2} \vec{e}_{\theta,\varphi} \hat{\Gamma}_{\theta,\varphi} u_{\theta,\varphi}(t) \sin\theta d\theta d\varphi, \tag{3.60}$$

where

$$\hat{\Gamma}_{\theta,\varphi} u_{\theta,\varphi}(t) = \iint_{\alpha \geq \beta} \nu(\alpha, \beta, \theta, \varphi) \hat{\gamma}_{\alpha\beta} u_{\theta,\varphi}(t) d\alpha d\beta. \tag{3.61}$$

The last two formulas can be combined into one expression:

$$\vec{f}(t) = \int_0^{2\pi} \int_0^{\pi/2} \vec{e}_{\theta,\varphi} \left(\iint_{\alpha \geq \beta} \nu(\alpha, \beta, \theta, \varphi) \hat{\gamma}_{\alpha\beta} u_{\theta,\varphi}(t) d\alpha d\beta \right) \sin\theta d\theta d\varphi. \tag{3.62}$$

Similarly, the 2-D vector Preisach model (3.55) can be represented in polar coordinates as

$$\vec{f}(t) = \int_{-\pi/2}^{\pi/2} \vec{e}_\varphi \hat{\Gamma}_\varphi u_\varphi(t) d\varphi, \tag{3.63}$$

where

$$\hat{\Gamma}_\varphi u_\varphi(t) = \iint_{\alpha \geq \beta} \nu(\alpha, \beta, \varphi) \hat{\gamma}_{\alpha\beta} u_\varphi(t) d\alpha d\beta. \tag{3.64}$$

In the previous formulas, $\vec{e}_\varphi$ is a unit vector along the direction specified by a polar angle φ, $\hat{\Gamma}_\varphi$ is the scalar Preisach model (operator) for this direction, and $u_\varphi(t)$ is the projection of $\vec{u}(t)$ along the direction of $\vec{e}_\varphi$.

By combining (3.63) and (3.65), we have

$$\vec{f}(t) = \int_{-\pi/2}^{\pi/2} \vec{e}_\varphi \left(\iint_{\alpha \geq \beta} \nu(\alpha, \beta, \varphi) \hat{\gamma}_{\alpha\beta} u_\varphi(t) d\alpha d\beta \right) d\varphi. \tag{3.65}$$

The expressions (3.62) and (3.65) are written for anisotropic models. In the isotropic case, the function ν should be independent of θ and φ (or only of φ in 2-D case). This leads to the following 3-D and 2-D isotropic vector Preisach models

$$\vec{f}(t) = \int_0^{2\pi} \int_0^{\pi/2} \vec{e}_{\theta,\varphi} \left(\iint_{\alpha \geq \beta} \nu(\alpha, \beta) \hat{\gamma}_{\alpha\beta} u_{\theta,\varphi}(t) d\alpha d\beta \right) \sin \theta d\theta d\varphi, \tag{3.66}$$

$$\vec{f}(t) = \int_{-\pi/2}^{\pi/2} \vec{e}_\varphi \left(\iint_{\alpha \geq \beta} \nu(\alpha, \beta) \hat{\gamma}_{\alpha\beta} u_\varphi(t) d\alpha d\beta \right) d\varphi. \tag{3.67}$$

In the models described above, the functions ν have not yet been specified. These functions should be determined by fitting the vector models to some experimental data. This is an identification problem. It is apparent that the identification problem is the central one as far as practical applications of the above vector hysteresis models are concerned. This problem will be discussed in sufficient detail in the subsequent sections. However, it is appropriate to comment already here that the solution of the identification problem is significantly simplified by the introduction of the auxiliary function $P(\alpha, \beta, \theta, \varphi)$. For any fixed θ and φ, consider a triangle $T(\alpha, \beta)$ shown in Fig. 3.10. Then by definition, we have

$$P(\alpha, \beta, \theta, \varphi) = \iint_{T(\alpha,\beta)} \nu(\alpha', \beta', \theta, \varphi) d\alpha' d\beta'. \tag{3.68}$$

By using (3.68), it can easily be shown that P is related to ν by the formula

$$\nu(\alpha, \beta, \theta, \varphi) = -\frac{\partial^2 P(\alpha, \beta, \theta, \varphi)}{\partial\alpha \partial\beta}. \tag{3.69}$$

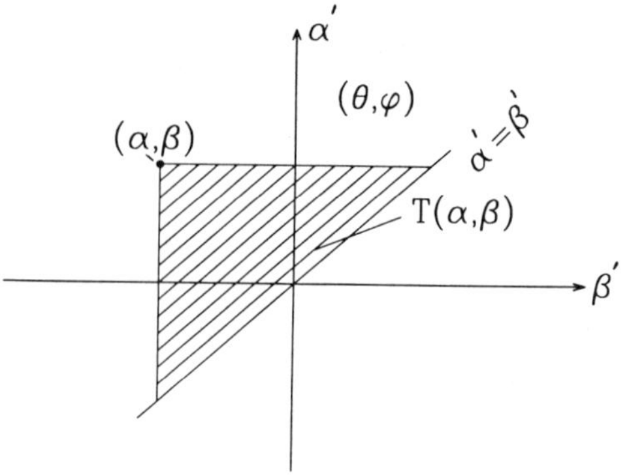

Fig. 3.10

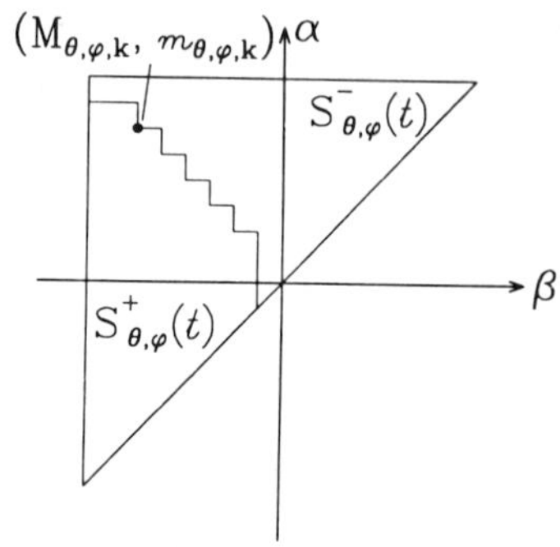

Fig. 3.11

Thus, if the function P is somehow determined, then the function ν can be easily retrieved. However, from the computational point of view, it is more convenient to use the function P than ν. This is because the double integrals with respect to α and β in expressions (3.62), (3.65), (3.66), and (3.67) can be explicitly expressed in terms of P, and in this way the above double integration can be completely avoided. Indeed, for any fixed direction $\vec{e}_{\theta,\varphi}$ we can consider the corresponding $\alpha - \beta$ diagram. A typical example of such a diagram is shown in Fig. 3.11, where $M_{\theta,\varphi,k}$ and $m_{\theta,\varphi,k}$ form an alternating series of past dominant maxima and minima of input projections along the direction specified by $\vec{e}_{\theta,\varphi}$. By using these maxima and minima, the output of the scalar Preisach model $\hat{\Gamma}_{\theta,\varphi}\vec{u}_{\theta,\varphi}(t)$ associated with the direction $\vec{e}_{\theta,\varphi}$ can be evaluated as

$$\hat{\Gamma}_{\theta,\varphi}u_{\theta,\varphi}(t)$$
$$= -P(\alpha_0, \beta_0) + 2\sum_{k=1}^{n_{\theta,\varphi}(t)} [P(M_{\theta,\varphi,k}, m_{\theta,\varphi,k-1}, \theta, \varphi) - P(M_{\theta,\varphi,k}, m_{\theta,\varphi,k}, \theta, \varphi)]. \tag{3.70}$$

The proof of (3.70) literally repeats the proof of the expression (1.69) for the classical Preisach model, and, for this reason, it is omitted.

In the case of 2-D vector Preisach model (3.63)–(3.65), a similar expression is valid for $\hat{\Gamma}_\varphi u_\varphi(t)$:

$$\hat{\Gamma}_\varphi u_\varphi(t) = -P(\alpha_0, \beta_0) + 2\sum_{k=1}^{n_\varphi(t)} [P(M_{\varphi,k}, m_{\varphi,k-1}, \varphi) - P(M_{\varphi,k}, m_{\varphi,k}, \varphi)]. \tag{3.71}$$

By using formulas (3.70) and (3.71) the numerical implementation of 3-D vector model (3.60)–(3.62) and 2-D vector model (3.63)–(3.65) can be reduced to the evaluation of double and single integrals, respectively. We note here that the numerical implementation of the classical 2-D Stoner-Wohlfarth model (3.32) requires the evaluation of double integrals. In this respect, numerical implementation of a 2-D vector Preisach model can be accomplished more efficiently than the numerical implementation of the 2-D Stoner-Wohlfarth model. Another advantage of using the formulas (3.70) and (3.71) is that the function P can be directly related to experimental data. This will be demonstrated when we study the identification problems for the vector Preisach models.

Using the expressions (3.70) and (3.71), digital codes which implement the vector Preisach models (3.60)–(3.62) and (3.63)–(3.65) have been developed. In these codes finite meshes of directions $\vec{e}_{\theta,\varphi}$ and $\vec{e}_{\varphi}$ are used to evaluate double and single integrals in (3.60) and (3.63), respectively. For each direction from these meshes, the integrands in (3.60) and (3.63) are computed by using (3.70) and (3.71). Some numerical examples computed by using the developed digital codes will be given in the next section.

3 Some Basic Properties of Vector Preisach Hysteresis Models

In the previous section we have defined the vector Preisach models of hysteresis and discussed their numerical implementation. The purpose of this section is to study some basic properties of these models and to show that these properties are qualitatively similar to those observed in experiments.

We begin with the property of reduction of vector hysteresis to scalar hysteresis. It has been mentioned before that this property is experimentally observed when an input is restricted to vary along an arbitrary chosen direction. We shall show that a similar reduction property holds for the vector Preisach models; this property is stated more precisely below. For the sake of notational simplicity it is formulated and proven only for the 2-D Preisach model (3.63)–(3.65), although it holds for the 3-D Preisach model (3.60)–(3.62) as well.

Reduction Property of the Vector Preisach Model to the Scalar Preisach Model

Consider an input $\vec{u}(t)$ restricted to vary along some direction $\vec{e}_{\varphi_0}$ for times $t \geq t_0$. Suppose that during $t \geq t_0$, $u_{\varphi_0}(t) = u(t)$ consecutively reaches values u_+ and u_- (with $u_+ > u_-$) and remains thereafter within these bounds. Then, for the Preisach vector model (3.63)–(3.65), *the relationship between the output projection $f_{\varphi_0}(t)$ along the direction $\vec{e}_{\varphi_0}$ and the input $u(t)$ exhibits the wiping-out and congruency properties. Since these properties constitute necessary and sufficient conditions for the representation of hysteresis nonlinearities by the classical scalar Preisach model, we conclude that the vector Preisach model is reduced to the scalar Preisach model. If u_+ and u_- are finite, this reduction is limited; if u_- and u_+ are equal to plus and minus infinity, respectively, the above reduction is complete.*

Proof

Without impairing the generality of our discussion, we can assume that $\varphi_0 = 0$. Then, for any φ the input projection $u_\varphi(t)$ varies between $u_+ \cos\varphi$ and $u_- \cos\varphi$. This means that for any φ input variations may affect $\alpha - \beta$ diagrams only within the triangle $T(u_+ \cos\varphi, u_- \cos\varphi)$ (see Fig. 3.12). It is also clear that all input projections $u_\varphi(t)$ reach maximum or minimum values at the same time and that these extremum values are "cos φ-multiples" of the corresponding extremum values of $u_0(t) = u(t)$. Consequently, if $\{M_k\}$ and $\{m_k\}$ constitute an alternating series of past dominant extrema of $u(t)$, then $\{M_{\varphi,k}\}$ and $\{m_{\varphi,k}\}$ constitute a corresponding alternating series of dominant extrema of $u_\varphi(t)$ under the condition that

$$M_{\varphi,k} = M_k \cos\varphi, \qquad m_{\varphi,k} = m_k \cos\varphi. \tag{3.72}$$

From (3.63), we find

$$f_x(t) = \int_{-\pi/2}^{\pi/2} \cos\varphi \hat{\Gamma}_\varphi u_\varphi(t) d\varphi. \tag{3.73}$$

From (3.71), (3.72) and (3.73) we conclude that only the alternating series of dominant extrema M_k and m_k of $u(t)$ affect the value of $f_x(t)$. All other input extrema are wiped out. This is tantamount to the wiping-out property for the hysteretic relation between $f_x(t)$ and $u(t)$.

We shall next prove that the above hysteretic relation also exhibits congruency of minor loops. Let $u^{(1)}(t)$ and $u^{(2)}(t)$ be two inputs which vary between u_+ and u_- for $t \geq t_0$ and which may have different past histories for $t_0 \leq t \leq t_0'$. However, starting from instant of time t_0', these inputs vary back and forth between the same two consecutive extrema u_+' and u_-'. As a result of these back-and-forth input variations, some minor loops are formed. We intend to show that these minor loops are congruent. The proof of the congruency of the above loop is equivalent to showing that any equal increments of inputs $u^{(1)}(t)$ and $u^{(2)}(t)$ result in equal increments of outputs $f_x^{(1)}(t)$ and $f_x^{(2)}(t)$. To this end, let us assume that both inputs after achieving the same value u_-' are increased by the same amount: $\Delta u^{(1)} = \Delta u^{(2)} = \Delta u$. As a result of these in-

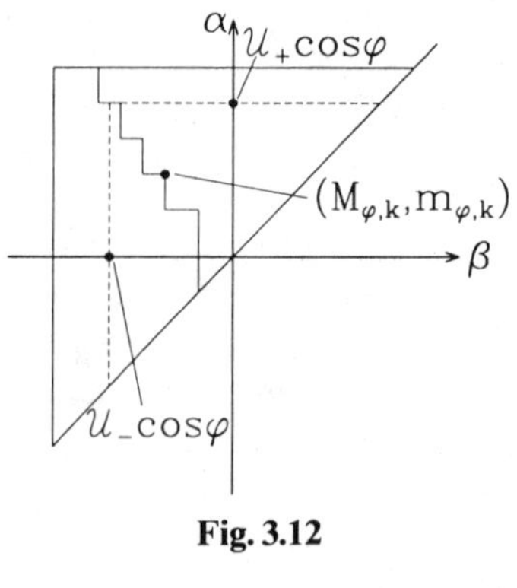

Fig. 3.12

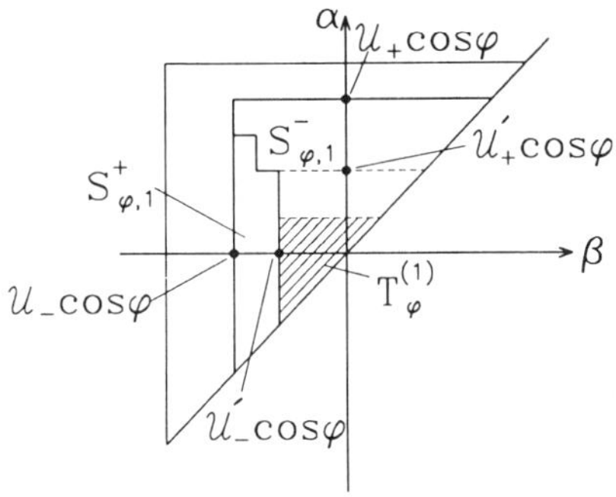

Fig. 3.13

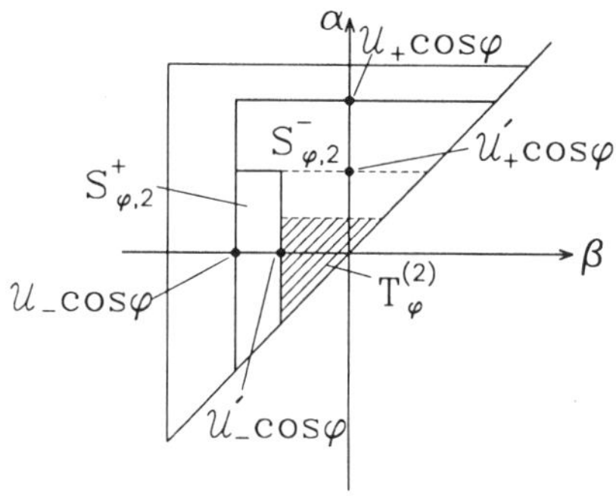

Fig. 3.14

creases, the identical triangles $T_\varphi^{(1)}$ and $T_\varphi^{(2)}$ are added to positive sets $S_{\varphi,1}^+(t)$ and $S_{\varphi,2}^+(t)$ and subtracted from the negative sets $S_{\varphi,1}^-(t)$ and $S_{\varphi,x}^-(t)$ (see Fig. 3.13 and 3.14). Since

$$\hat{\Gamma}_\varphi u_\varphi(t) = \iint\limits_{S_\varphi^+(t)} \nu(\alpha, \beta, \varphi)\, d\alpha\, d\beta - \iint\limits_{S_\varphi^-(t)} \nu(\alpha, \beta, \varphi)\, d\alpha\, d\beta, \tag{3.74}$$

from (3.73) we derive

$$\Delta f_x^{(1)} = 2 \int_{-\pi/2}^{\pi/2} \cos\varphi \left(\iint\limits_{T_\varphi^{(1)}} \nu(\alpha, \beta, \varphi)\, d\alpha\, d\beta \right) d\varphi, \tag{3.75}$$

$$\Delta f_x^{(2)} = 2 \int_{-\pi/2}^{\pi/2} \cos\varphi \left(\iint\limits_{T_\varphi^{(2)}} \nu(\alpha, \beta, \varphi)\, d\alpha\, d\beta \right) d\varphi. \tag{3.76}$$

Since $T_\varphi^{(1)} = T_\varphi^{(2)}$ for any φ, we conclude that

$$\Delta f_x^{(1)} = \Delta f_x^{(2)}. \tag{3.77}$$

The equality (3.77) has been proven for the case when inputs $u^{(1)}(t)$ and $u^{(2)}(t)$ are monotonically increased by the same amount after achieving the same minimum value u'_-. Thus, this equality means the congruency for the ascending branches of the above minor loops. By literally repeating the previous reasoning, we can prove that the same equality (3.77) holds when the inputs $u^{(1)}(t)$ and $u^{(2)}(t)$ are monotonically decreased by the same amount Δu after achieving the same maximum value u'_+. This implies the congruency of descending branches of the above minor loops. Thus, the congruency property for minor loops is established. This completes the proof of the validity of the reduction property.

In our discussion of the reduction property, we have proven the congruency of "scalar" minor loops described by the vector Preisach models. The last result admits the following generalization.

Congruency Property of Vector Minor Loops

Let the tips of two inputs $\vec{u}^{(1)}(t)$ and $\vec{u}^{(2)}(t)$ trace the same closed curve for $t \geq t_0$ (see Fig. 3.15). Then the tips of the corresponding outputs $\vec{f}^{(1)}(t)$ and $\vec{f}^{(2)}(t)$ of the vector Preisach models trace congruent closed curves for $t \geq t_0$ (see Fig. 3.16). These curves may be noncollocated in space because of possibly different input histories prior to t_0.

Proof

Consider the 3-D Preisach model (3.60)–(3.62). Since the tips of both inputs $\vec{u}^{(1)}(t)$ and $\vec{u}^{(2)}(t)$ trace the same closed curve, we conclude that for any direction $\vec{e}_{\theta,\varphi}$ the corresponding $\alpha - \beta$ diagrams are modified with time in the identical way within the same triangles $T(u_{+,\theta,\varphi}, u_{-,\theta,\varphi})$. For instance, as the tips of the inputs $\vec{u}^{(1)}(t)$ and $\vec{u}^{(2)}(t)$ move from the point $\vec{a}$ to the point b (see Fig. 3.15), the same regions $\Omega_{1,\theta,\varphi}$ and $\Omega_{2,\theta,\varphi}$ may be added to the positive sets $S^+_{\theta,\varphi,1}$ and $S^+_{\theta,\varphi,2}$ and subtracted from the negative sets $S^-_{\theta,\varphi 1}$ and $S^-_{\theta,\varphi,2}$. Since

$$\hat{\Gamma}_{\theta,\varphi} u_{\theta,\varphi}(t) = \iint\limits_{S^+_{\theta,\varphi}(t)} \nu(\alpha, \beta, \theta, \varphi)\, d\alpha\, d\beta - \iint\limits_{S^-_{\theta,\varphi}(t)} \nu(\alpha, \beta, \theta, \varphi)\, d\alpha\, d\beta, \tag{3.78}$$

we conclude that

$$\Delta\hat{\Gamma}_{\theta,\varphi} u^{(1)}_{\theta,\varphi}(t) = 2 \iint\limits_{\Omega_{1,\theta,\varphi}} \nu(\alpha, \beta, \theta, \varphi)\, d\alpha\, d\beta, \tag{3.79}$$

$$\Delta\hat{\Gamma}_{\theta,\varphi} u^{(2)}_{\theta,\varphi}(t) = 2 \iint\limits_{\Omega_{2,\theta,\varphi}} \nu(\alpha, \beta, \theta, \varphi)\, d\alpha\, d\beta. \tag{3.80}$$

By using (3.60), (3.79), and (3.80), we find that the corresponding output increments which connect the points $A^{(1)}$, $B^{(1)}$ and $A^{(2)}$, $B^{(2)}$ (see Fig. 3.16) are given by

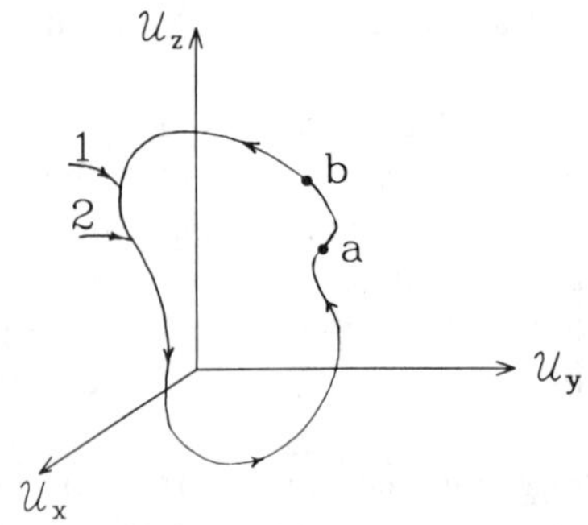

Fig. 3.15

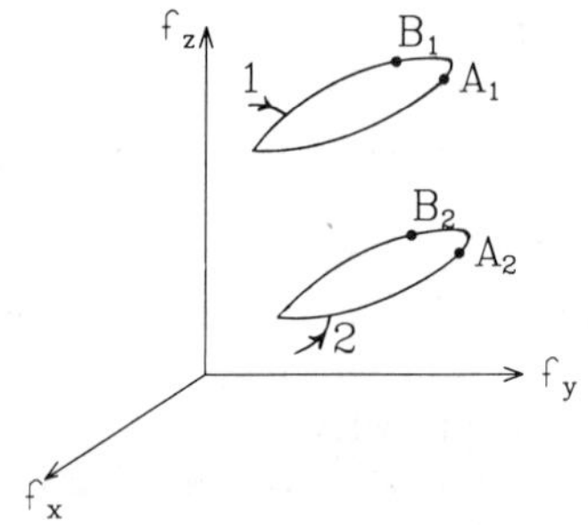

Fig. 3.16

$$\Delta\vec{f}^{(1)} = 2\int_0^{2\pi}\int_0^{\pi} \vec{e}_{\theta,\varphi}\left(\iint\limits_{\Omega_{1,\theta,\varphi}} \nu(\alpha, \beta, \theta, \varphi)d\alpha d\beta\right)\sin\theta d\theta d\varphi, \quad (3.81)$$

$$\Delta\vec{f}^{(2)} = 2\int_0^{2\pi}\int_0^{\pi} \vec{e}_{\theta,\varphi}\left(\iint\limits_{\Omega_{2,\theta,\varphi}} \nu(\alpha, \beta, \theta, \varphi)d\alpha d\beta\right)\sin\theta d\theta d\varphi. \quad (3.82)$$

Since $\Omega_{1,\theta,\varphi} = \Omega_{2,\theta,\varphi}$ for any θ and φ, from (3.81) and (3.82), we conclude that

$$\Delta\vec{f}^{(1)} = \Delta\vec{f}^{(2)}. \quad (3.83)$$

The equality (3.83) holds for arbitrarily chosen points a and b, and this proves that the vector minor loops shown in Fig. 3.16 are congruent.

We next proceed to the discussion of one remarkable property which is valid for the 2-D isotropic vector Preisach model (3.67).

Rotational Symmetry Property

Consider a uniformly rotating input (that is, one of constant magnitude and angular velocity):

$$\vec{u}(t) = \{u_x(t) = u_m \cos\omega t,\ u_y = u_m \sin\omega t\}. \quad (3.84)$$

Then the output of the 2-D *isotropic Preisach model* (3.67) *can be represented as*

$$\vec{f}(t) = \vec{f}_0 + \vec{\tilde{f}}(t), \quad (3.85)$$

where $\vec{f}_0$ *does not change with time, while* $\vec{\tilde{f}}(t)$ *is a uniformly rotating vector.*

Proof

It is clear from the very definition of the uniformly rotating input $\vec{u}(t)$ and Fig. 3.17 that

$$u_\varphi(t) = u_m \cos(\omega t - \varphi). \quad (3.86)$$

According to the theorem proven in Section 2 of this chapter, the integration over any semicircle can be used in the definition of the 2-D vector Preisach model. This fact allows one to modify the definition (3.67) of the 2-D isotropic Preisach model as follows:

$$\vec{f}(t) = \frac{1}{2}\int_0^{2\pi} \vec{e}_\varphi\left(\iint\limits_{\alpha\geq\beta} \nu(\alpha, \beta)\hat{\gamma}_{\alpha\beta}u_\varphi(t)d\alpha d\beta\right)d\varphi. \quad (3.87)$$

It is clear from (3.86) that for all directions $\vec{e}_\varphi$ the corresponding $\alpha - \beta$ diagrams are modified with time within the same triangles $\tilde{T} = T(u_m, -u_m)$. Outside of these triangles, the $\alpha - \beta$ diagrams remain unchanged. These unchanged parts of $\alpha - \beta$ diagrams contribute to the term $\vec{f}_0$ in (3.85), while the time-varying parts of $\alpha - \beta$ diagrams result in the time-varying term $\tilde{\vec{f}}(t)$. By using the above comment as well as (3.86) and (3.87), we find

$$\tilde{\vec{f}}(t) = \frac{1}{2}\int_0^{2\pi} \vec{e}_\varphi \left(\iint_{\tilde{T}} \nu(\alpha, \beta)\hat{\gamma}_{\alpha\beta} u_m \cos(\omega t - \varphi) d\alpha d\beta \right) d\varphi. \tag{3.88}$$

The expression (3.88) can be represented in terms of cartesian components as

$$\tilde{f}_x(t) = \frac{1}{2}\int_0^{2\pi} \cos\varphi \left(\iint_{\tilde{T}} \nu(\alpha, \beta)\hat{\gamma}_{\alpha\beta} u_m \cos(\omega t - \varphi) d\alpha d\beta \right) d\varphi, \tag{3.89}$$

$$\tilde{f}_y(t) = \frac{1}{2}\int_0^{2\pi} \sin\varphi \left(\iint_{\tilde{T}} \nu(\alpha, \beta)\hat{\gamma}_{\alpha\beta} u_m \cos(\omega t - \varphi) d\alpha d\beta \right) d\varphi. \tag{3.90}$$

Consider some instant of time t. For this instant of time, all directions $\vec{e}_\varphi$ can be divided into two sets such that

$$0 \le \omega t - \varphi \le \pi, \tag{3.91}$$

and

$$\pi \le \omega t - \varphi \le 2\pi. \tag{3.92}$$

For the first set, all input projections $u_\varphi(t)$ are monotonically decreasing, and this results in the $\alpha - \beta$ diagram shown in Fig. 3.18. For the second set, all input projections $u_\varphi(t)$ are monotonically increasing, and this results in the $\alpha - \beta$ diagram shown in Fig. 3.19. Next we introduce a new variable:

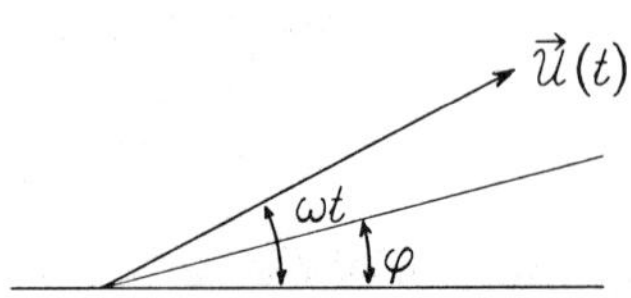

Fig. 3.17

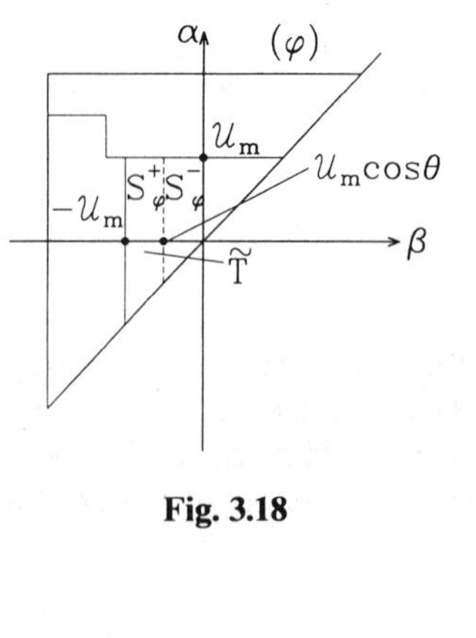

Fig. 3.18

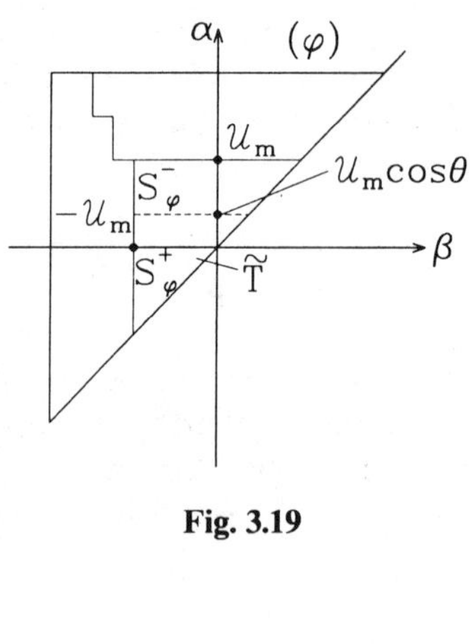

Fig. 3.19

$$\theta = \omega t - \varphi. \tag{3.93}$$

It is apparent that the double integral over $\tilde{T}$ in (3.89) and (3.90) is the function of θ. This justifies the notation

$$G(\theta) = \iint_{\tilde{T}} \nu(\alpha, \beta)\hat{\gamma}_{\alpha\beta} u_m \cos(\omega t - \varphi) d\alpha d\beta. \tag{3.94}$$

If

$$0 \leq \theta \leq \pi, \tag{3.95}$$

then the diagram shown in Fig. 3.18 is valid, and from this diagram and formulas (3.68) and (3.94) we find

$$G(\theta) = P(u_m, -u_m) - 2P(u_m, u_m \cos \theta). \tag{3.96}$$

If

$$\pi \leq \theta \leq 2\pi, \tag{3.97}$$

then the diagram shown in Fig. 3.19 is valid, and as before we find

$$G(\theta) = 2P(u_m \cos \theta, -u_m) - P(u_m, -u_m). \tag{3.98}$$

By using the change of variables (3.93), the notation (3.94), and by taking into account that

$$d\theta = -d\varphi, \tag{3.99}$$

and

$$\omega t - 2\pi \leq \theta \leq \omega t, \tag{3.100}$$

we transform (3.89) as follows:

$$\vec{f}_x(t) = \frac{1}{2}\int_0^{2\pi} \cos(\omega t - \theta) G(\theta) d\theta = A \cos \omega t + B \sin \omega t, \tag{3.101}$$

where

$$A = \frac{1}{2}\int_0^{2\pi} \cos \theta G(\theta) d\theta, \tag{3.102}$$

$$B = \frac{1}{2}\int_0^{2\pi} \sin \theta G(\theta) d\theta. \tag{3.103}$$

Similarly, the expression (3.90) can be reduced to the form

$$\tilde{f}_y(t) = A \sin \omega t - B \cos \omega t. \tag{3.104}$$

The formulas (3.101) and (3.104) can be modified to

$$\tilde{f}_x(t) = \sqrt{A^2 + B^2} \cos(\omega t - \xi), \tag{3.105}$$

$$\tilde{f}_y(t) = \sqrt{A^2 + B^2} \sin(\omega t - \xi), \tag{3.106}$$

where

$$\tan \xi = \frac{B}{A}. \tag{3.107}$$

From (3.105) and (3.106), we conclude that $\vec{\vec{f}}$ is a uniformly rotating vector with the magnitude equal to $\sqrt{A^2 + B^2}$. We next express A and B in terms of the function P and prove that the angle ξ in (3.105) and (3.106) is acute. From (3.95)–(3.98) and (3.102), we find

$$A = \frac{1}{2}\left\{\int_0^{\pi} \cos\theta[P(u_m, -u_m) - 2P(u_m, u_m\cos\theta)]d\theta \right.$$
$$\left. + \int_{\pi}^{2\pi} \cos\theta[2P(u_m\cos\theta, -u_m) - P(u_m, -u_m)]d\theta\right\}. \tag{3.108}$$

In the second integral in (3.108), we will use the change of variables

$$\theta' = \theta - \pi \tag{3.109}$$

and also take into account that

$$P(-u_m\cos\theta, -u_m) = P(u_m, u_m\cos\theta). \tag{3.110}$$

The last formula easily follows from the symmetry property (3.58) and the definition (3.68) of the function P. By using (3.109) and (3.110), we can transform (3.108) as follows:

$$A = -2\int_0^{\pi} \cos\theta P(u_m, u_m\cos\theta)d\theta. \tag{3.111}$$

From (3.95)–(3.98) and (3.103), we obtain

$$B = \frac{1}{2}\left\{\int_0^{\pi} \sin\theta[P(u_m, -u_m) - 2P(u_m, u_m\cos\theta)]d\theta \right.$$
$$\left. + \int_{\pi}^{2\pi} \sin\theta[2P(u_m\cos\theta, -u_m) - P(u_m, -u_m)]d\theta\right\}. \tag{3.112}$$

To transform the second integral in (3.112), we will use the change of variables (3.109) and the identity (3.110). This eventually leads to the following expression for B:

$$B = 2P(u_m, -u_m) - 2\int_0^{\pi} \sin\theta P(u_m, u_m\cos\theta)\, d\theta. \tag{3.113}$$

Thus, we have found explicit expressions (3.111) and (3.113) for A and B in terms of the function P. In the next section, it will be shown that P can be related to some experimental data. In this way we can relate A and B to the experimental data, and, consequently, find the magnitude and phase of the uniformly rotating vector $\vec{\vec{f}}(t)$ in terms of these data.

Now, we will use the formulas (3.111) and (3.113) to prove that the phase angle ξ is acute under some general conditions. The first condition can be expressed mathematically as

$$P(\alpha, \beta) \leq P(\alpha, \beta') \quad \text{if} \quad \beta \geq \beta'. \tag{3.114}$$

The condition (3.114) means that $P(\alpha, \beta)$ is a monotonically decreasing function of β for any fixed α. It is clear from the definition (3.68) of P that the condition (3.114) is satisfied if $\nu(\alpha, \beta)$ is positive.

From (3.111), we find

$$A = -2\left[\int_0^{\pi/2} \cos\theta P(u_m, u_m \cos\theta)d\theta + \int_{\pi/2}^{\pi} \cos\theta P(u_m, u_m \cos\theta)d\theta\right]. \tag{3.115}$$

By using the change of variables $\theta' = \pi - \theta$ in the second integral in (3.115), we derive

$$A = -2\int_0^{\pi/2} \cos\theta[P(u_m, u_m \cos\theta) - P(u_m, -u_m \cos\theta)]d\theta. \tag{3.116}$$

From (3.114) and (3.116), we conclude

$$A > 0. \tag{3.117}$$

The second condition is

$$\int_0^{\pi} \sin\theta[P(u_m, -u_m) - 2P(u_m, u_m \cos\theta)]d\theta > 0. \tag{3.118}$$

From (3.113) and (3.118), we find

$$B > 0. \tag{3.119}$$

From (3.107), (3.117), and (3.119), we conclude that the phase angle ξ is acute. This completes the discussion of the rotational symmetry property.

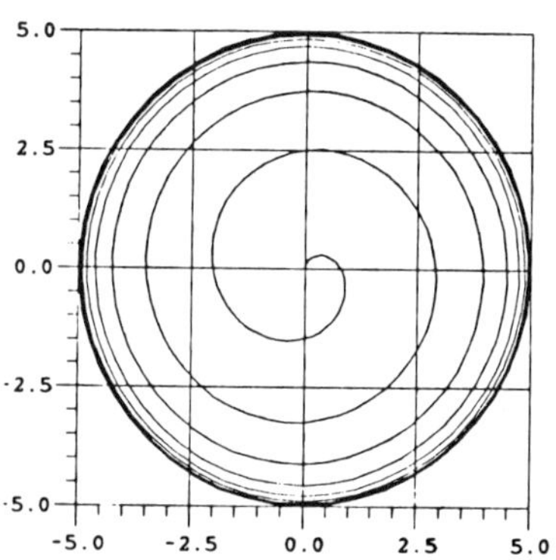

Fig. 3.20

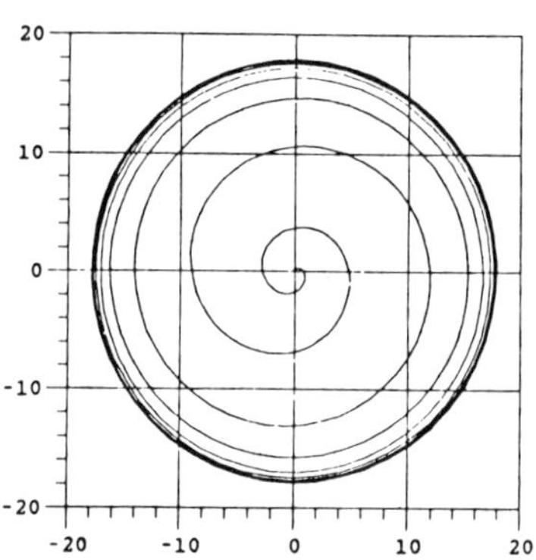

Fig. 3.21

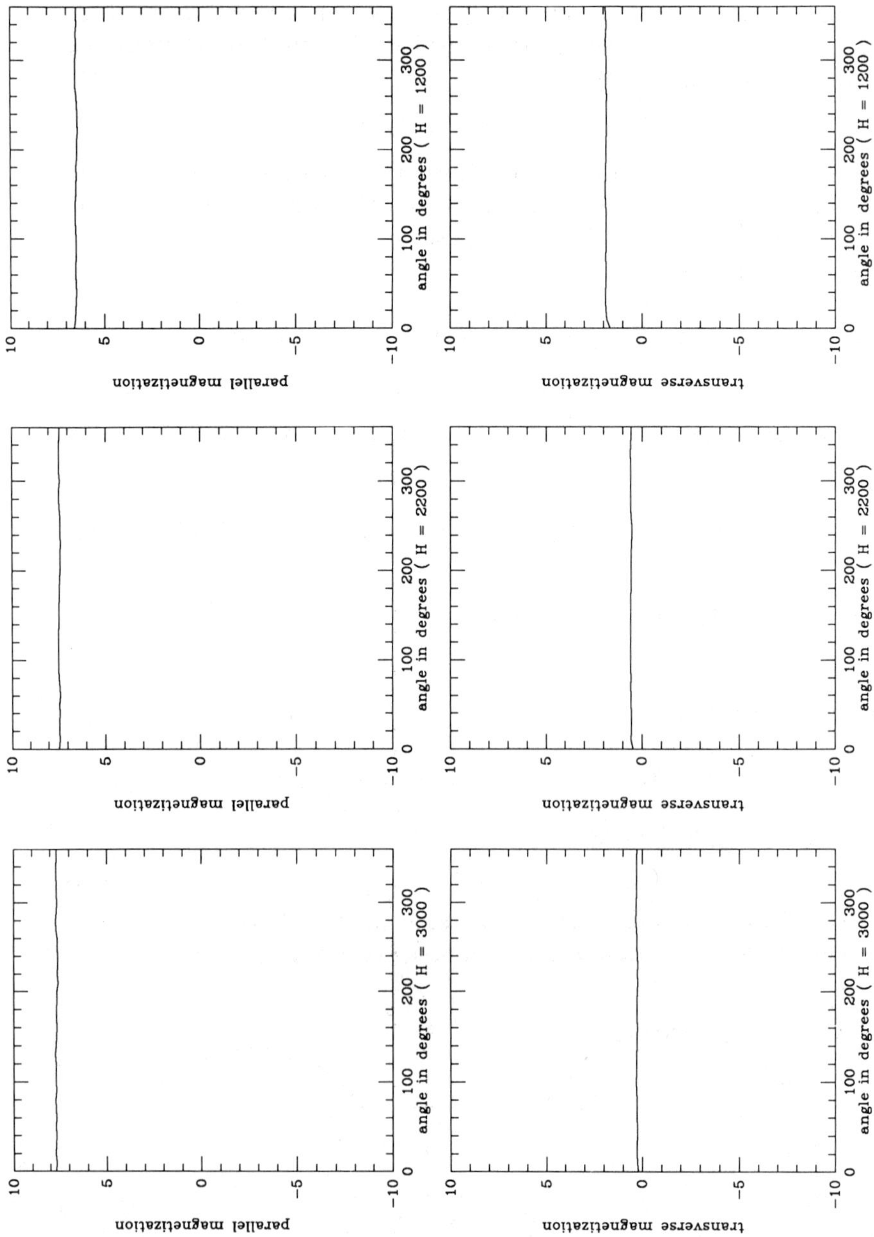
parallel magnetization
transverse magnetization
angle in degrees (H = 1200)
angle in degrees (H = 2200)
angle in degrees (H = 3000)
10
5
0
-5
-10
0
100
200
300

Figure 3.22a

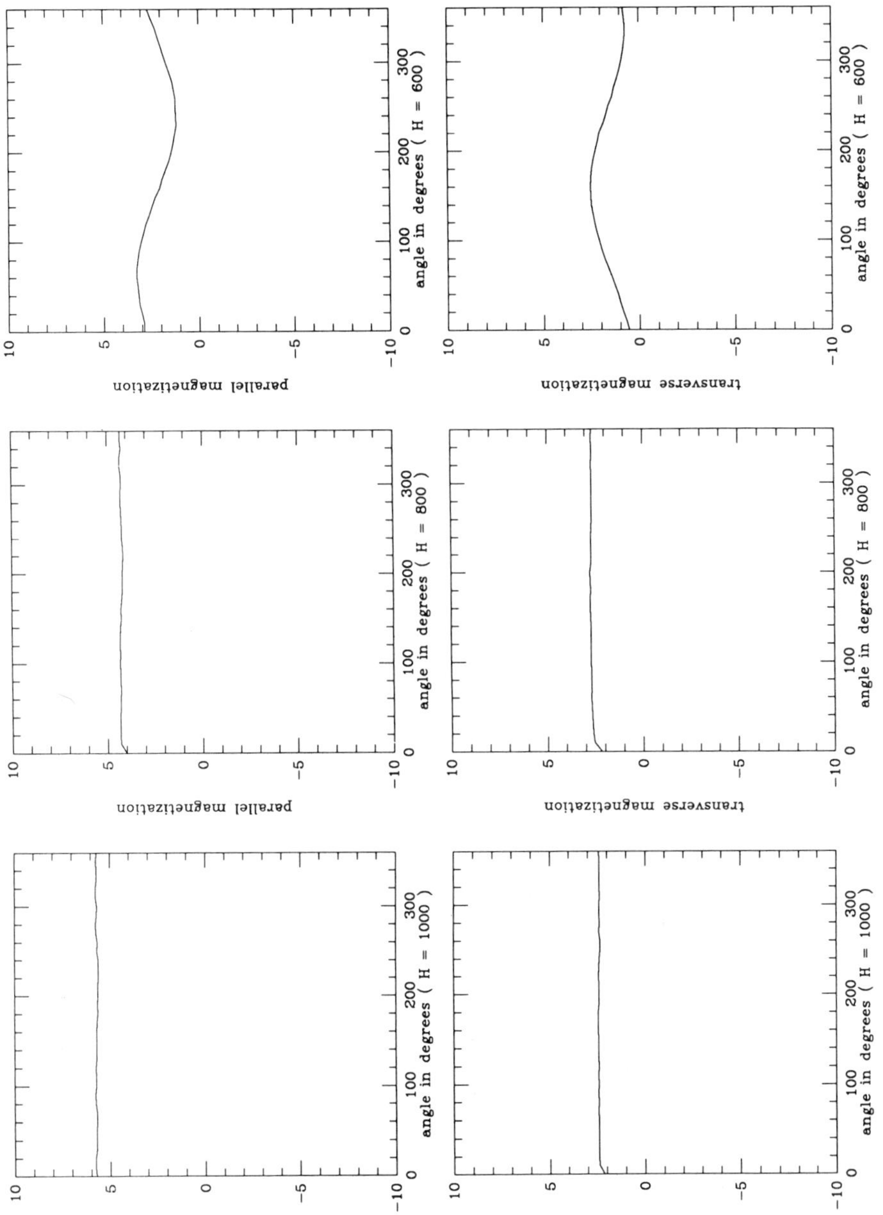
parallel magnetization
transverse magnetization
angle in degrees (H = 600)
angle in degrees (H = 800)
angle in degrees (H = 1000)

Figure 3.22b

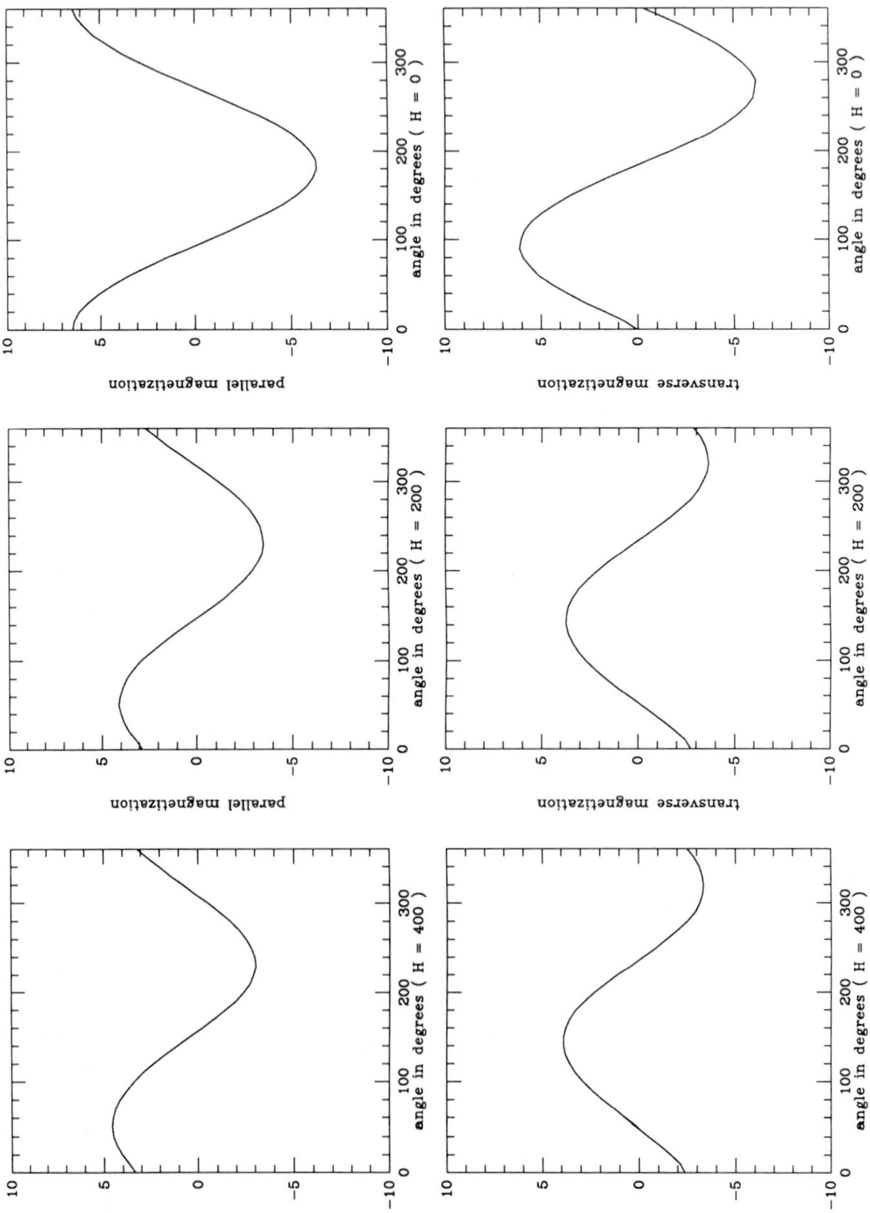
parallel magnetization
angle in degrees (H = 0)
transverse magnetization
angle in degrees (H = 0)
parallel magnetization
angle in degrees (H = 200)
transverse magnetization
angle in degrees (H = 200)
parallel magnetization
angle in degrees (H = 400)
transverse magnetization
angle in degrees (H = 400)

Figure 3.22c

It is worthwhile noting that in spite of the nonlinear structure of the vector Preisach model (3.67), the time-harmonic input (3.84) produces (up to a history-dependent constant term $\vec{f}_0$) a time-harmonic output. In other words, no generation of higher-order harmonics is caused by the nonlinear structure of the vector Preisach model. This remarkable fact admits the following physical explanation. The isotropic vector Preisach model (3.67) has a mathematical form which is invariant with respect to any rotation of cartesian coordinates. The mathematical form of uniformly rotating input (3.84) is also invariant with respect to any rotation of cartesian coordinates. Thus, on the physical grounds we expect that the mathematical form of the resulting output should be also invariant with respect to any rotation of cartesian coordinates. This is possible only if the output is a uniformly rotating vector. The above discussion clearly reveals the meaning of the term "rotational symmetry property."

The rotational symmetry property has been confirmed by numerical computations. By using a digital code which implements the vector Preisach model (3.67) and has briefly been described in the previous section, the output of the Preisach model has been computed for the input shown in Fig. 3.20. This input gradually approaches the regime of uniform rotation (3.84). The results of computations shown in Fig. 3.21 demonstrate that the output also gradually attains the regime of uniform rotation. The property of rotational symmetry has also been observed in the experiments. The results of these experiments are shown in Fig. 3.22. In these experiments, the applied magnetic fields were stationary, while the magnetizable sample of Isomax material was uniformly rotated. Thus, in these experiments, the magnetization was measured in the coordinate frame uniformly rotating with respect to the sample. The component of magnetization measured along the applied field is called "parallel magnetization," while the component perpendicular to the field is named "transverse magnetization." The results shown in Fig. 3.22 clearly suggest that for large applied fields the whole magnetization of the sample moves in synchronism with the uniformly rotating coordinate frame. This means that we deal with uniformly rotating magnetization. As the applied field is reduced, the time constant term of magnetization appears which reveals itself (in the uniformly rotating frame) as sinusoidally changing components of magnetization.

The described experiments were performed by my students on the premises of the Naval Air Development Center (Warminster, PA). The help and valuable advices of Dr. C. Krafft concerning the performance of the above experiments are gratefully acknowledged.

Property of Correlation Between Mutually Orthogonal Components of Output and Input

Suppose that the input $\vec{u}(t)$ was first restricted to vary along the y-axis. It was increased from infinitely negative value to some positive value u_+, and then it was decreased to zero. Some remanent value of output $\vec{f}_r = \vec{e}_y f_r$ resulted

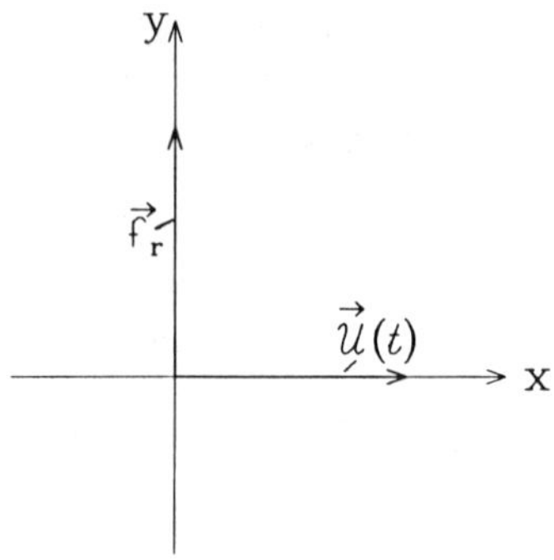

Fig. 3.23

from the above input variations. After reaching zero, the input is restricted to vary along the orthogonal x-direction (see Fig. 3.23). It is asserted that, by increasing the input in x-direction, it is possible to reduce the orthogonal remanent component of the output to any, however small, value.

Proof

All directions $\vec{e}_\varphi$ can be subdivided into sets:

$$0 \leq \varphi \leq \frac{\pi}{2} \quad \text{and} \quad -\frac{\pi}{2} \leq \varphi \leq 0. \tag{3.120}$$

The input variations along the y-axis result in $\alpha - \beta$ diagrams shown in Figs. 3.24 and 3.25 for the first and second sets of directions, respectively. As the input is increased along the x-axis, the above diagrams are modified. These modifications are shown by dash lines. They result in continuous expansions of positive set $S_\varphi^+(t)$ at the expense of negative set $S_\varphi^-(t)$.

From (3.67), we find

$$f_y(t) = \int_{-\pi/2}^{\pi/2} \sin\varphi \left(\iint\limits_{\alpha \geq \beta} \nu(\alpha, \beta) \hat{\gamma}_{\alpha\beta} u_x(t) \cos\varphi \, d\alpha \, d\beta \right) d\varphi. \tag{3.121}$$

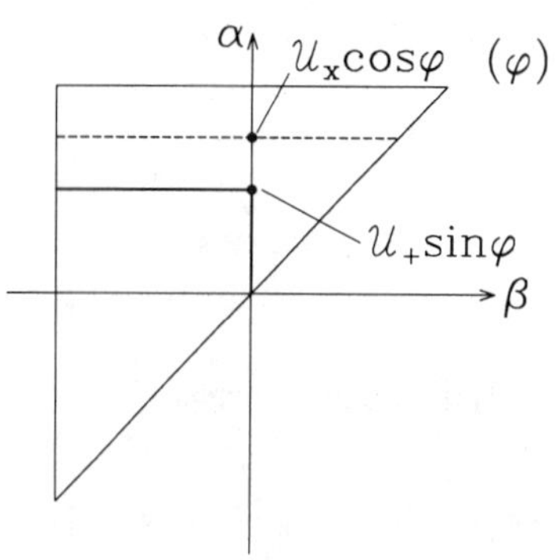

Fig. 3.24

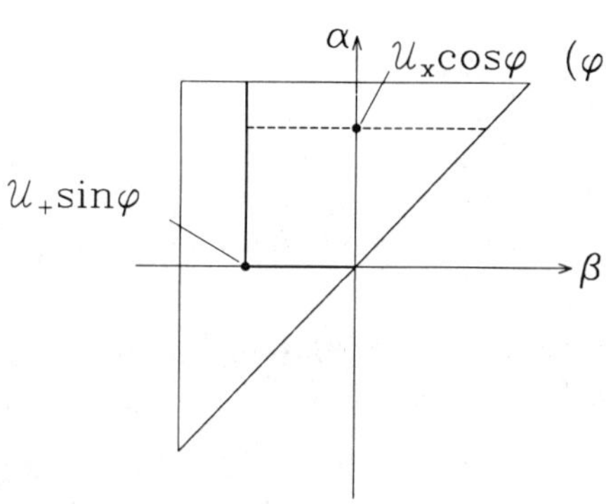

Fig. 3.25

By introducing the notation

$$G(\varphi, t) = \iint_{\alpha \geq \beta} \nu(\alpha, \beta)\hat{\gamma}_{\alpha\beta}u_x(t)\cos\varphi\, d\alpha\, d\beta, \tag{3.122}$$

we can represent (3.121) in the form

$$f_y(t) = \int_0^{\pi/2} \sin\varphi G(\varphi, t)d\varphi + \int_{-\pi/2}^{0} \sin\varphi G(\varphi, t)d\varphi. \tag{3.123}$$

By changing φ to $-\varphi$ in the second integral in (3.123), we obtain

$$f_y(t) = \int_0^{\pi/2} \sin\varphi[G(\varphi, t) - G(-\varphi, t)]d\varphi. \tag{3.124}$$

If $u_x \cos\varphi \geq \alpha_0$, then according to the diagrams shown in Figs. 3.24 and 3.25 we have

$$G(\varphi, t) = G(-\varphi, t). \tag{3.125}$$

From (3.124) and (3.125), we find

$$f_y(t) = \int_{\arccos(\alpha_0/u_x(t))}^{\pi/2} \sin\varphi[G(\varphi, t) - G(-\varphi, t)]d\varphi. \tag{3.126}$$

Since $\arccos(\alpha_0/u_x(t)) \to \pi/2$ as $u_x(t) \to \infty$, we conclude that

$$f_y(t) \to 0. \tag{3.127}$$

This completes the proof of the above property.

This property has been confirmed by numerical computations performed by using a digital code which implements the model (3.67). The results of computations are shown in Fig. 3.26 for different remanent values of the output. The property of correlation between orthogonal components of input

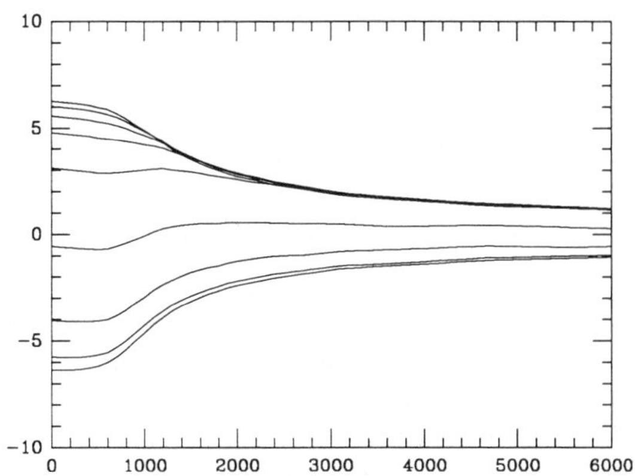

Fig. 3.26

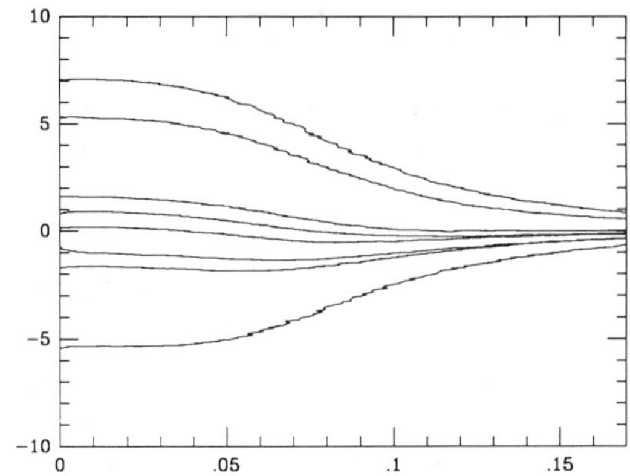

Fig. 3.27

and output has also been observed in the experiments. The results of these experiments are presented in Fig. 3.27. There is apparent qualitative similarity between the computational and experimental results. It is worthwhile noting here that the above property of orthogonal correlation has been regarded as an important "testing" property for any vector hysteresis models in magnetics.

4 Identification Problem for Isotropic Vector Preisach Models

The essence of the identification problem is in determining the function $\nu(\alpha, \beta)$ or $P(\alpha, \beta)$ from some experimental data. It turns out that this problem can be reduced to the solution of a special integral equation which relates the function $P(\alpha, \beta)$ to some "unidirectional" (scalar) hysteresis data. We first present the derivation of this equation for the 2-D model (3.67). The derivation proceeds as follows.

Consider the projection $f_x(t)$ of $\vec{f}(t)$ along the direction $\varphi = 0$. According to the equation (3.67), we have

$$f_x(t) = \int_{-\pi/2}^{\pi/2} \cos\varphi \left(\iint_{\alpha' \geq \beta'} \nu(\alpha', \beta') \hat{\gamma}_{\alpha'\beta'} u_\varphi(t) d\alpha' d\beta' \right) d\varphi. \tag{3.128}$$

Now we restrict $\vec{u}(t)$ to vary along the direction $\varphi = 0$ which means that $\vec{u}(t) = \vec{e}_x u(t)$. First, we assume that $u(t)$ is made "infinitely negative." Then for any α', β' and φ we have $\hat{\gamma}_{\alpha'\beta'} u_\varphi(t) = -1$. Next, we assume that the input is monotonically increased until it reaches some value α. Let f_α denote the corresponding value of $f_x(t)$. As a result of the above input increase, we find that for any φ we have $\hat{\gamma}_{\alpha'\beta'} u_\varphi(t) = +1$ if $\alpha' < \alpha \cos\varphi$ and $\hat{\gamma}_{\alpha'\beta'} u_\varphi(t) = -1$ if $\alpha' > \alpha \cos\varphi$. This is shown geometrically in Fig. 3.28. Consequently, for any

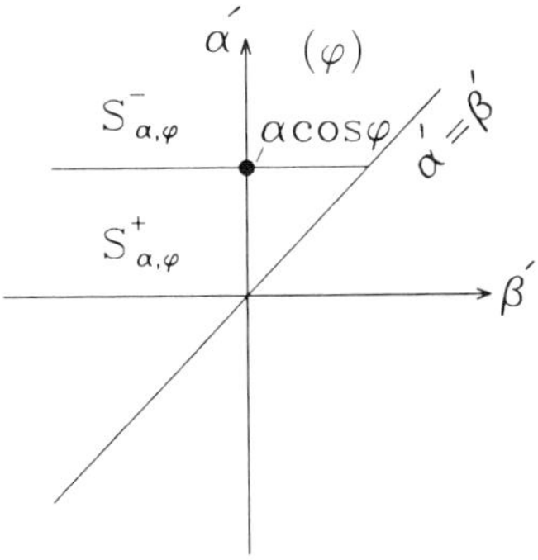

Fig. 3.28

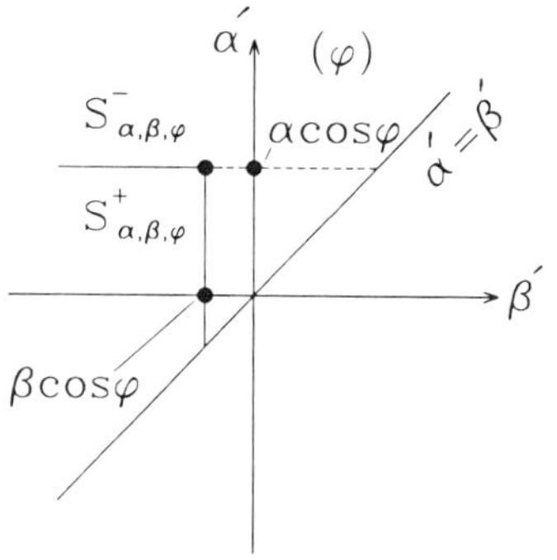

Fig. 3.29

φ we have

$$\iint_{\alpha' \geq \beta'} \nu(\alpha', \beta')\hat{\gamma}_{\alpha'\beta'} u_\varphi(t) d\alpha' d\beta' = \iint_{S^+_{\alpha,\varphi}} \nu(\alpha', \beta') d\alpha' d\beta' - \iint_{S^-_{\alpha,\varphi}} \nu(\alpha', \beta') d\alpha' d\beta'. \tag{3.129}$$

Finally, we assume that the above input increase is followed by a subsequent monotonic decrease of $u(t)$ until it reaches some value β. Let $f_{\alpha\beta}$ denote the corresponding value of $f_x(t)$. The above input decrease modifies the previous geometric diagram which now for any φ assumes the form shown in Fig. 3.29. According to this figure, we have

$$\iint_{\alpha' \geq \beta'} \nu(\alpha', \beta')\hat{\gamma}_{\alpha'\beta'} u_\varphi(t) d\alpha' d\beta' = \iint_{S^+_{\alpha,\beta,\varphi}} \nu(\alpha', \beta') d\alpha' d\beta' - \iint_{S^-_{\alpha,\beta,\varphi}} \nu(\alpha', \beta') d\alpha' d\beta'. \tag{3.130}$$

Consider the function

$$F(\alpha, \beta) = \tfrac{1}{2}(f_\alpha - f_{\alpha\beta}). \tag{3.131}$$

By using formulas (3.68) and (3.128)–(3.131), it is straightforward to show that $P(\alpha, \beta)$ is related to $F(\alpha, \beta)$ by the expression

$$\int_{-\pi/2}^{\pi/2} \cos\varphi P(\alpha\cos\varphi, \beta\cos\varphi) d\varphi = F(\alpha, \beta). \tag{3.132}$$

It is apparent that f_α and $f_{\alpha\beta}$ can be found by measuring the first-order transition curves. Of course, the input $u(t)$ can not be made infinitely negative as was assumed in the derivation. However, by making the input "sufficiently negative," a reasonable accuracy can be secured.

The formula (3.132) can be construed as the integral equation which relates $P(\alpha, \beta)$ to the first-order transition curves. This integral equation has a peculiar structure which is revealed by the following result.

Theorem

Consider the operator

$$\hat{A}P = \int_{-\pi/2}^{\pi/2} \cos\varphi P(\alpha\cos\varphi, \beta\cos\varphi)\,d\varphi. \tag{3.133}$$

Monomials $\alpha^k\beta^s$ are eigenfunctions of the operator $\hat{A}$.

Proof

From (3.133) we find

$$\begin{aligned}\hat{A}\alpha^k\beta^s &= \int_{-\pi/2}^{\pi/2} \cos\varphi\,\alpha^k\cos^k\varphi\,\beta^s\cos^s\varphi\,d\varphi \\ &= \alpha^k\beta^s \int_{-\pi/2}^{\pi/2} \cos^{k+s+1}\varphi\,d\varphi = \lambda_{k+1}\alpha^k\beta^s,\end{aligned} \tag{3.134}$$

where

$$\lambda_{k+s} = \begin{cases} \pi\dfrac{(2n-1)!!}{(2n)!!}, & \text{if } k+s+1 = 2n, \\[2ex] 2\dfrac{(2n)!!}{(2n+1)!!}, & \text{if } k+s+1 = 2n+1. \end{cases} \tag{3.135}$$

By using the above theorem, the solution of equation (3.132) can be easily found when the right-hand side $F(\alpha, \beta)$ is a polynomial:

$$F(\alpha, \beta) = \sum_{m=0}^{M} \sum_{k+s=m} a_{ks}^{(m)}\alpha^k\beta^s. \tag{3.136}$$

Indeed, looking for the solution of the equation in the form

$$P(\alpha, \beta) = \sum_{m=0}^{M} \sum_{k+s=m} P_{ks}^{(m)}\alpha^k\beta^s, \tag{3.137}$$

from the above theorem we find

$$P_{ks}^{(m)} = \frac{a_{ks}^{(m)}}{\lambda_m}. \tag{3.138}$$

By using this technique, the following algorithm for the solution of the equation (3.132) can be suggested. We first extend the function $F(\alpha, \beta)$ from the triangle $-\alpha_0 \le \beta \le \alpha \le \alpha_0$ to the square $-\alpha_0 \le \alpha \le \alpha_0$, $-\alpha_0 \le \beta \le \alpha_0$, for instance, as an even function with respect to the diagonal $\alpha = \beta$. Then we expand this function into the series of Chebyshev polynomials

$$F(\alpha, \beta) = \sum_{\ell, q} a_{\ell q} T_\ell(\alpha) T_q(\beta), \tag{3.139}$$

where the Chebyshev polynomials are defined as

$$T_0(\alpha) = 1, \quad T_\ell(\alpha) = \frac{1}{2^{\ell-1}} \cos(\ell \text{ arc cos } \alpha). \tag{3.140}$$

By assuming for the sake of notational simplicity that $\alpha_0 = 1$, for the expansion coefficients $a_{\ell q}$ in (3.139) we have the formula (see [62])

$$a_{\ell q} = \frac{2^{\ell+q}}{\pi^2} \int_{-1}^{1} \int_{-1}^{1} \frac{F(\alpha, \beta) T_\ell(\alpha) T_q(\beta)}{\sqrt{1-\alpha^2}\sqrt{1-\beta^2}} d\alpha d\beta. \tag{3.141}$$

Let $t_{\ell q}(\alpha, \beta)$ be the solution of the integral equations

$$\int_{-\pi/2}^{\pi/2} \cos \varphi t_{\ell q}(\alpha \cos \varphi, \beta \cos \varphi) d\varphi = T_\ell(\alpha) T_q(\beta), \tag{3.142}$$

then the solution of the integral equation (3.132) can be represented in the form

$$P(\alpha, \beta) = \sum_{\ell, q} a_{\ell q} t_{\ell q}(\alpha, \beta). \tag{3.143}$$

Thus, there are two major steps in finding the solution (3.143) of the integral equation (3.132). The first step is to solve the equations (3.142). This can be accomplished by using the technique (3.136)–(3.138). The second step is to evaluate the expansion coefficients $a_{\ell q}$. This can be handled by using the formula (2.141). However, for the sake of computations it is desirable to transform this formula as follows. By introducing new variables θ and θ':

$$\alpha = \cos \theta, \qquad \beta = \cos \theta'; \tag{3.144}$$

we find

$$T_\ell(\alpha) = \frac{\cos \ell\theta}{2^{\ell-1}}, \qquad T_q(\beta) = \frac{\cos q\theta'}{2^{q-1}}, \tag{3.145}$$

$$a_{\ell q} = \frac{4}{\pi^2} \int_0^{\pi} \int_0^{\pi} F(\cos \theta, \cos \theta') \cos \ell\theta \cos q\theta' d\theta d\theta'. \tag{3.146}$$

Thus, the expansion coefficients $a_{\ell q}$ can be evaluated by using fast cosine-transform schemes.

There is another approach to the solution of the integral equation (3.132) which leads to a closed form expression for $P(\alpha, \beta)$. This approach is based on the following change of variables:

$$x = \alpha \cos \varphi, \qquad \lambda = \beta/\alpha, \tag{3.147}$$

which after simple transformations yields the following integral equation of the Abel type:

$$\int_0^{\alpha} \frac{x}{\sqrt{\alpha^2 - x^2}} P(x, \lambda x) dx = \frac{\alpha}{2} F(\alpha, \lambda\alpha). \tag{3.148}$$

Some simplification of the above equation is achieved by introducing new auxiliary functions

$$N(x) = xP(x, \lambda x), \qquad R(\alpha) = \frac{\alpha}{2} F(\alpha, \lambda \alpha). \tag{3.149}$$

Then, from (3.148) we obtain

$$\int_0^\alpha \frac{N(x)dx}{\sqrt{\alpha^2 - x^2}} = R(\alpha). \tag{3.150}$$

The solution of the above equation can be found by using the following trick. We multiply both sides of (3.150) by $\alpha/\sqrt{s^2 - \alpha^2}$ and integrate with respect to α from 0 to s:

$$\int_0^s \frac{\alpha}{\sqrt{s^2 - \alpha^2}} \left(\int_0^\alpha \frac{N(x)dx}{\sqrt{\alpha^2 - x^2}} \right) d\alpha = \int_0^s \frac{\alpha R(\alpha)}{\sqrt{s^2 - \alpha^2}} d\alpha. \tag{3.151}$$

By using Fig. 3.30, we find

$$\begin{aligned} \int_0^s \frac{\alpha}{\sqrt{s^2 - \alpha^2}} \left(\int_0^\alpha \frac{N(x)dx}{\sqrt{\alpha^2 - x^2}} \right) d\alpha &= \iint_\Omega \frac{\alpha N(x) dx d\alpha}{\sqrt{(s^2 - \alpha^2)(\alpha^2 - x^2)}} \\ &= \int_0^s N(x) \left(\int_x^s \frac{\alpha d\alpha}{\sqrt{(s^2 - \alpha^2)(\alpha^2 - x^2)}} \right) dx. \end{aligned} \tag{3.152}$$

Next, by using simple transformations, we obtain

$$\int_x^s \frac{\alpha d\alpha}{\sqrt{(s^2 - \alpha^2)(\alpha^2 - x^2)}} = \frac{1}{2} \int_x^s \frac{d(\alpha^2)}{\sqrt{\left(\frac{s^2 - x^2}{2}\right)^2 - \left(\alpha^2 - \frac{s^2 + x^2}{2}\right)^2}}. \tag{3.153}$$

By introducing a new variable of integration

$$\omega = \alpha^2 - \frac{s^2 + x^2}{2}, \tag{3.154}$$

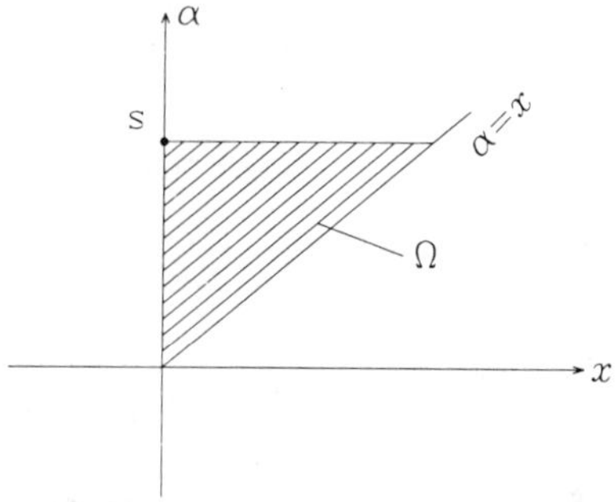

Fig. 3.30

from (3.153) we derive

$$\int_x^s \frac{\alpha d\alpha}{\sqrt{(s^2 - x^2)(\alpha^2 - x^2)}} = \frac{1}{2}\int_{(x^2-s^2)/2}^{(s^2-x^2)/2} \frac{d\omega}{\sqrt{\left(\frac{s^2 - x^2}{2}\right)^2 - \omega^2}}$$

$$= \frac{1}{2}\arcsin\frac{2\omega}{s^2 - x^2}\Bigg|_{\omega=(x^2-s^2)/2}^{\omega=(s^2-x^2)/2} = \frac{\pi}{2}. \qquad (3.155)$$

From (3.151), (3.152) and (3.155), we obtain

$$\int_0^s N(x)dx = \frac{2}{\pi}\int_0^s \frac{\alpha R(\alpha)}{\sqrt{s^2 - \alpha^2}} d\alpha. \qquad (3.156)$$

By differentiating (3.156) with respect to s and then by replacing s by α and α by s, we find

$$N(\alpha) = \frac{2}{\pi}\frac{d}{d\alpha}\int_0^\alpha \frac{sR(s)}{\sqrt{\alpha^2 - s^2}} ds. \qquad (3.157)$$

This is the closed form solution to the integral equation (3.150). Now, by substituting (3.149) into (3.157), we find the closed form expression for P:

$$P(\alpha, \lambda\alpha) = \frac{1}{\pi\alpha}\frac{d}{d\alpha}\int_0^\alpha \frac{s^2 F(s, \lambda s)}{\sqrt{\alpha^2 - s^2}} ds. \qquad (3.158)$$

The last expression is valid for any value of λ. By varying λ from 0 to 1 and using (3.158), we can compute all required values of P. However, for actual computations, it is convenient to transform (3.158) by integration by parts. The final expression is then given by

$$P(\alpha, \lambda\alpha) = \frac{1}{\pi}\int_0^\alpha \frac{F(s, \lambda s) + s\frac{d}{ds}F(s, \lambda s)}{\sqrt{\alpha^2 - s^2}} ds. \qquad (3.159)$$

Now, we turn to the three-dimensional isotropic model (3.66). Consider the projection $f_z(t)$ of $\vec{f}(t)$ along the direction $\theta = 0$. According to (3.66), we have

$$f_z(t) = \int_0^{2\pi}\int_0^{\pi/2} \cos\theta \left[\iint_{\alpha' \geq \beta'} \nu(\alpha', \beta')\hat{\gamma}_{\alpha'\beta'}u_{\theta,\varphi}(t)d\alpha' d\beta'\right] \sin\theta d\theta d\varphi. \quad (3.160)$$

We restrict $\vec{u}(t)$ to vary along the direction $\theta = 0$, which means that $\vec{u}(t) = \vec{e}_z u(t)$. As before, we first assume that $u(t)$ is made infinitely negative. Then for any α, β, θ and φ we have $\hat{\gamma}_{\alpha\beta}u_\theta$, $\varphi(t) = -1$. Next we assume that the input is monotonically increased until it reaches some value α. Let f_α denote the corresponding value of $f_z(t)$. As a result of the above input increase, we find that for any θ and φ we have $\hat{\gamma}_{\alpha'\beta'}u_{\theta,\varphi}(t) = +1$ if $\alpha' < \alpha\cos\theta$ and $\hat{\gamma}_{\alpha'\beta'}u_{\theta,\varphi}(t) =$

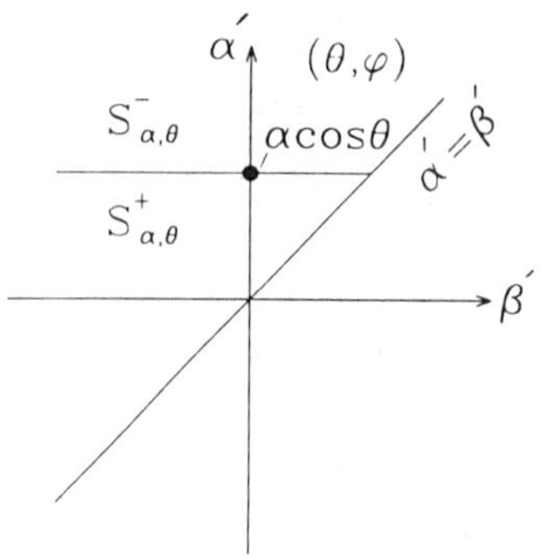

Fig. 3.31

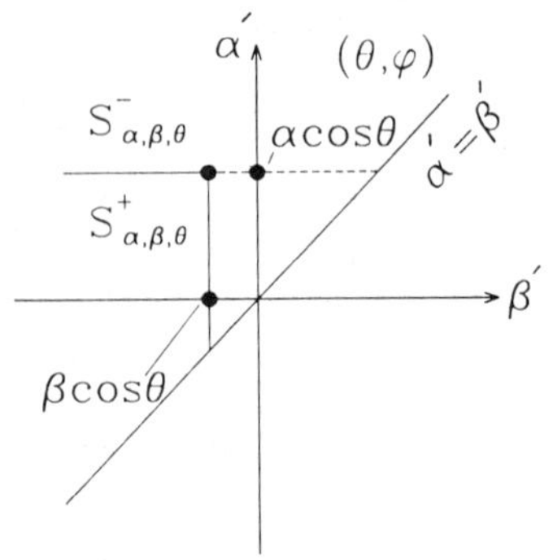

Fig. 3.32

-1 if $\alpha' > \alpha \cos \theta$. This is illustrated geometrically in Fig. 3.31. Consequently, for any θ and φ we have

$$\iint\limits_{\alpha' \geq \beta'} \nu(\alpha', \beta')\hat{\gamma}_{\alpha'\beta'} u_{\theta,\varphi}(t)d\alpha' d\beta' = \iint\limits_{S^+_{\alpha,\theta}} \nu(\alpha', \beta')d\alpha' d\beta' - \iint\limits_{S^-_{\alpha,\theta}} \nu(\alpha', \beta')d\alpha' d\beta'. \tag{3.161}$$

Finally, we assume that the above input increase is followed by a subsequent monotonic decrease of $u(t)$ until it reaches some value β. Let $f_{\alpha\beta}$ denote the corresponding value of $f_z(t)$. The above-mentioned input decrease results in geometric diagrams shown in Fig. 3.32. According to this figure, we have

$$\iint\limits_{\alpha' \geq \beta'} \nu(\alpha', \beta')\hat{\gamma}_{\alpha'\beta'}, u_{\theta,\varphi}(t)d\alpha' d\beta' = \iint\limits_{S^+_{\alpha,\beta,\theta}} \nu(\alpha', \beta')d\alpha' d\beta' - \iint\limits_{S^-_{\alpha,\beta,\theta}} \nu(\alpha', \beta')d\alpha' d\beta'. \tag{3.162}$$

By substituting (3.161) and (3.162) into (3.160), then by subtracting one expression from another and by using (3.68) and (3.131), we derive

$$\int_0^{2\pi} \int_0^{\pi/2} \cos \theta P(\alpha \cos \theta, \beta \cos \theta)\sin \theta d\theta d\varphi = F(\alpha, \beta). \tag{3.163}$$

Since the integrand does not depend on φ, we obtain

$$\int_0^{\pi/2} \cos \theta \sin \theta P(\alpha \cos \theta, \beta \cos \theta)d\theta = \frac{1}{2\pi} F(\alpha, \beta). \tag{3.164}$$

This is the integral equation which relates $P(\alpha, \beta)$ to the experimentally measured first-order transition curves. It turns out that this equation is easily solvable. By using the change of variables

$$x = \alpha \cos \theta, \qquad \lambda = \beta/\alpha, \tag{3.165}$$

the equation (3.164) can be represented in the form

$$\int_0^{\alpha} xP(x, \lambda x)dx = \frac{\alpha^2}{2\pi} F(\alpha, \lambda\alpha). \tag{3.166}$$

By differentiating the last equation with respect to α, we find

$$\alpha P(\alpha, \lambda\alpha) = \frac{1}{2\pi}\frac{d}{d\alpha}[\alpha^2 F(\alpha, \lambda\alpha)]. \tag{3.167}$$

which leads to the following final expression:

$$P(\alpha, \lambda\alpha) = \frac{1}{2\pi\alpha}\frac{d}{d\alpha}[\alpha^2 F(\alpha, \lambda\alpha)]. \tag{3.168}$$

It is remarkable that the solution of the identification problem for 3-D isotropic Preisach models of vector hysteresis turns out to be much simpler than the solution of the same problem for 2-D isotropic models.

5 Identification Problem for Anisotropic Vector Preisach Models

We shall first discuss the identification problem for the two-dimensional anisotropic model (3.65). As before, we shall use the function $P(\alpha, \beta, \varphi)$ defined by (3.68). We shall relate this function to the experimental data represented by the sets of first-order transition curves measured along all directions $\vec{e}_\varphi$. These experimental data can be characterized by the function

$$F(\alpha, \beta, \varphi) = \tfrac{1}{2}(f_{\alpha\varphi} - f_{\alpha\beta\varphi}), \tag{3.169}$$

which will be assumed to be known in the subsequent discussion. Of course, it is not feasible to measure first-order transition curves for all directions $\vec{e}_\varphi$. However, it is possible to measure these curves for some finite meshes of directions $\vec{e}_\varphi$ and to use subsequent interpolation for computing $F(\alpha, \beta, \varphi)$ for all φ.

Consider local polar coordinates (ρ, ψ) with polar axis directed along the vector $\vec{e}_{\varphi'}$. The model (3.65) in the local polar coordinates can be written as

$$\vec{f}(t) = \int_{-\pi/2}^{\pi/2} \vec{e}_\psi \left(\iint_{\alpha \geq \beta} \nu(\alpha, \beta, \varphi' + \psi)\hat{\gamma}_{\alpha\beta} u_\psi(t) d\alpha d\beta \right) d\psi, \tag{3.170}$$

where the relationship between the angles φ, φ' and ψ is illustrated by Fig. 3.33.

By using the expression (3.170) and by repeating almost literally the same reasoning as in the derivation of integral equation (3.132), we find that the function $P(\alpha, \beta, \varphi)$ is related to the function $F(\alpha, \beta, \varphi)$ by the expression

$$\int_{-\pi/2}^{\pi/2} \cos\psi P(\alpha\cos\psi, \beta\cos\psi, \varphi' + \psi)d\psi = F(\alpha, \beta, \varphi'). \tag{3.171}$$

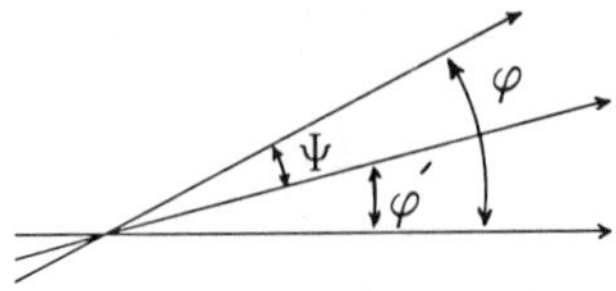

Fig. 3.33

Next we shall use the following Fourier expansions:

$$F(\alpha, \beta, \varphi') = \sum_{n=-\infty}^{\infty} F_n(\alpha, \beta)e^{in\varphi'}, \tag{3.172}$$

$$P(\alpha, \beta, \varphi) = \sum_{n=-\infty}^{\infty} P_n(\alpha, \beta)e^{in\varphi}, \tag{3.173}$$

$$\nu(\alpha, \beta, \varphi) = \sum_{n=-\infty}^{\infty} \nu_n(\alpha, \beta)e^{in\varphi}. \tag{3.174}$$

It is clear from (3.68) and (3.69) that P_n and ν_n are related by the expressions

$$P_n(\alpha, \beta) = \iint_{T(\alpha, \beta)} \nu_n(\alpha', \beta')d\alpha' d\beta', \tag{3.175}$$

$$\nu_n(\alpha, \beta) = -\frac{\partial^2 P_n(\alpha, \beta)}{\partial\alpha\partial\beta}. \tag{3.176}$$

By substituting (3.172) and (3.173) into (3.171), we find

$$\sum_{n=-\infty}^{\infty} e^{in\varphi'} \int_{-\pi/2}^{\pi/2} e^{in\psi} \cos\psi P_n(\alpha\cos\psi, \beta\cos\psi)d\psi = \sum_{n=-\infty}^{\infty} F_n(\alpha, \beta)e^{in\varphi'}. \tag{3.177}$$

From (3.177) we obtain

$$\int_{-\pi/2}^{\pi/2} e^{in\psi} \cos\psi P_n(\alpha\cos\psi, \beta\cos\psi)d\psi = F_n(\alpha, \beta), \quad (n = 0, \pm 1, \pm 2, \ldots). \tag{3.178}$$

Since $\sin n\psi \cos\psi P_n(\alpha\cos\psi, \beta\cos\psi)$ is an odd function of ψ, from (3.178) we derive

$$\int_{-\pi/2}^{\pi/2} \cos\psi \cos n\psi P_n(\alpha\cos\psi, \beta\cos\psi)d\psi = F_n(\alpha, \beta), \quad (n = 0, \pm 1, \pm 2, \ldots). \tag{3.179}$$

Thus we have obtained the infinite set of decoupled (separate) integral equations for P_n. The right-hand sides $F_n(\alpha, \beta)$ of these equations can be computed by using fast Fourier transform (FFT) algorithms.

For the case $n = 0$, from (3.179) we find

$$\int_{-\pi/2}^{\pi/2} \cos \psi P_0(\alpha \cos \psi, \beta \cos \psi) d\psi = F_0(\alpha, \beta), \tag{3.180}$$

which naturally coincides with the integral equation (3.132) for the isotropic vector Preisach model.

If $n = \pm 1$, from (3.179) we obtain

$$\int_{-\pi/2}^{\pi/2} \cos^2 \psi P_{\pm 1}(\alpha \cos \psi, \beta \cos \psi) d\psi = F_{\pm 1}(\alpha, \beta). \tag{3.181}$$

Closed form solutions to integral equations (3.180) and (3.181) can be found by using the change of variables

$$x = \alpha \cos \psi, \qquad \lambda = \beta/\alpha \tag{3.182}$$

and by reducing the above equations to the Abel-type integral equations. In the case of equation (3.180), the result is readily available and can be expressed by the formula

$$P_0(\alpha, \lambda\alpha) = \frac{1}{\pi\alpha} \frac{d}{d\alpha} \int_0^\alpha \frac{s^2 F_0(s, \lambda s)}{\sqrt{\alpha^2 - s^2}} ds. \tag{3.183}$$

In the case of equations (3.181), the change of variables (3.182) leads to the integral equations

$$\int_0^\alpha P_{\pm 1}(x, \lambda x) \frac{x^2 dx}{\sqrt{\alpha^2 - x^2}} = \frac{\alpha^2}{2} F_{\pm 1}(\alpha, \lambda\alpha). \tag{3.184}$$

These equations can be solved in exactly the same way as the equation (3.148). The final result is given by the following formula

$$P_{\pm 1}(\alpha, \lambda\alpha) = \frac{1}{\pi\alpha^2} \frac{d}{d\alpha} \int_0^\alpha \frac{s^3 F_{\pm 1}(s, \lambda s)}{\sqrt{\alpha^2 - s^2}} ds. \tag{3.185}$$

Thus, if we are interested only in the first three terms of the Fourier expansion for P (this is a first-order approximation for anisotropic media), then the solution of the identification problem can be found by using the closed form expressions (3.183) and (3.185). In the case of higher-order terms, the change of variables (3.182) leads to the following Abel-type integral equations

$$\int_0^\alpha P_n(x, \lambda x) \frac{x T_n(x/\alpha)}{\sqrt{\alpha^2 - x^2}} dx = \frac{\alpha}{2^n} F_n(\alpha, \lambda\alpha), \tag{3.186}$$

where T_n are Chebyshev polynomials defined by (3.190). Unfortunately, we have not been able to find closed-form solutions to these equations. However, there are many efficient numerical techniques developed for the solution of integral equations of this type (see, for instance, [64]).

The analytical form solutions to integral equations (3.179) can also be found by exploiting the following property of this equation.

Theorem

Consider operators

$$\hat{A}_n P = \int_{-\pi/2}^{\pi/2} \cos\psi \cos n\psi P(\alpha\cos\psi, \beta\cos\psi)d\psi. \tag{3.187}$$

Monomials $\alpha^k \beta^s$ are eigenfunctions of these operators.

Proof

It is straightforward to check the validity of the above statement and to get the following expressions for the eigenvalues $\lambda_{k+s}^{(n)}$:

$$\lambda_{k+s}^{(n)} = 2\int_0^{\pi/2} \cos n\psi \cos^{k+s+1}\psi d\psi. \tag{3.188}$$

The explicit formula for the integral in (3.188) can be found in the literature (see [62]).

By using the above theorem, polynomial solutions to the integral equations (3.179) can be easily found. Indeed, if the right-hand sides of these equations are polynomials in the form (3.136), then the solutions to these equations will be polynomials in the form (3.137). Polynomial coefficients of the right-hand sides and the solutions will be related by the expression (3.138) where instead of eigenvalues λ_m the eigenvalues $\lambda_{k+s}^{(n)}$ should be employed. By taking advantage of this fact, Chebyshev polynomial expansions of type (3.139) for $F_n(\alpha, \beta)$ can be used and the solutions can be found in the form (3.143), where the expansion coefficients can be computed by using expressions similar to (3.146).

Now, we proceed to the identification problem for the three-dimensional anisotropic vector Preisach model (3.62). This problem is technically more complicated than the corresponding problem for 2-D models, and, for this reason, it requires a special treatment. It turns out that some facts from the theory of irreducible representations of the group of rotations of three-dimensional Euclidian space are instrumental in the treatment of this identification problem.

As before we shall use the function $P(\alpha, \beta, \theta, \varphi)$ defined by (3.68), and we shall relate this function to the first-order transition curves experimentally measured along all possible directions $\vec{e}_{\theta,\varphi}$. These curves can be characterized by the function

$$F(\alpha, \beta, \theta, \varphi) = \tfrac{1}{2}(f_{\alpha\theta\varphi} - f_{\alpha\beta\theta\varphi}). \tag{3.189}$$

It will be convenient to use spherical harmonic expansions for the functions ν, P and F:

$$v(\alpha, \beta, \theta, \varphi) = \sum_{k=0}^{\infty} \sum_{m=-k}^{k} v_{km}(\alpha, \beta) Y_{km}(\theta, \varphi), \tag{3.190}$$

$$P(\alpha, \beta, \theta, \varphi) = \sum_{k=0}^{\infty} \sum_{m=-k}^{k} P_{km}(\alpha, \beta) Y_{km}(\theta, \varphi), \tag{3.191}$$

$$F(\alpha, \beta, \theta, \varphi) = \sum_{k=0}^{\infty} \sum_{m=-k}^{k} F_{km}(\alpha, \beta) Y_{km}(\theta, \varphi), \tag{3.192}$$

where $Y_{km}(\theta, \varphi)$ are spherical harmonics, while v_{km}, P_{km} and F_{km} are corresponding expansion coefficients. It is clear from (3.68) and (3.69) that P_{km} and v_{km} are related by the following expressions:

$$P_{km}(\alpha, \beta) = \iint_{T(\alpha,\beta)} v_{km}(\alpha', \beta') d\alpha' d\beta', \tag{3.193}$$

$$v_{km}(\alpha, \beta) = -\frac{\partial^2 P_{km}(\alpha, \beta)}{\partial\alpha\partial\beta}. \tag{3.194}$$

It turns out that the identification problem can be reduced to the solution of special integral equations which relate $P_{km}(\alpha, \beta)$ to $F_{km}(\alpha, \beta)$. The derivation of these equations proceeds as follows.

Consider an arbitrary direction specified by the angles θ' and φ'. We shall use a local coordinate system xyz which is obtained from the system XYZ by Euler rotations $R_Z(0)$, $R_X(\theta')$ and $R_Z(\pi/2 + \varphi')$ (see Fig. 3.34). The Euler angles for the inverse rotation from xyz to XYZ are equal to $\pi/2 - \varphi'$, θ' and π, respectively. It is clear that the direction of the axis z coincides with the direction of $\vec{e}_{\theta'\varphi'}$. By using local spherical coordinates ξ and ψ, we can represent the model (3.62) as

$$\vec{f}(t) = \int_0^{2\pi} \int_0^{\pi/2} \vec{e}_{\xi\psi} \left(\iint_{\alpha \geq \beta} v(\alpha, \beta, \xi, \psi, \theta', \varphi') \hat{\gamma}_{\alpha\beta} u_{\xi\psi}(t) d\alpha d\beta \right) \sin \xi d\xi d\psi. \tag{3.195}$$

By using the last expression and by repeating almost literally the reasoning

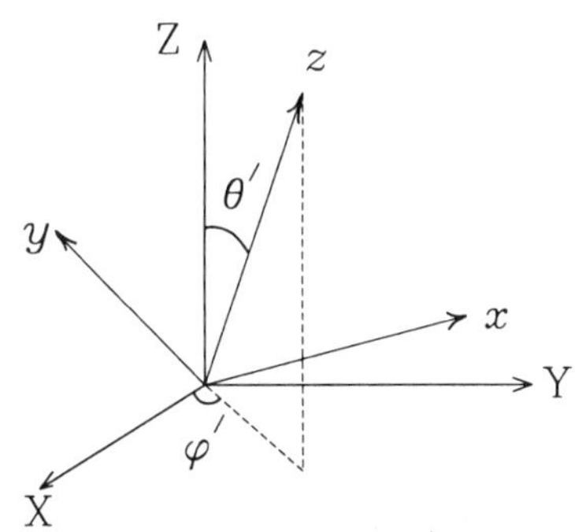

Fig. 3.34

which was used in the derivation of the integral equation (3.163), we can show that the model (3.195) matches first-order transition curves measured along the direction $\vec{e}_{\theta'\varphi'}$ if the function P satisfies the integral equation

$$\int_0^{2\pi}\int_0^{\pi/2} \cos\xi P(\alpha\cos\xi, \beta\cos\xi, \xi, \psi, \theta', \varphi')\sin\xi d\xi d\psi = F(\alpha, \beta, \theta', \varphi'). \tag{3.196}$$

Next, we shall relate the function P in local coordinates to the same function in spherical coordinates θ and φ. To achieve this, we shall use the spherical harmonic expansion (3.191) and the following facts from the theory of irreducible representations of the group of rotations (see [65], [66], [67]).

Linear combinations of spherical harmonics Y_{km} of the fixed order k form a linear space H_k of the irreducible representation of the group of rotations. This means that any spherical harmonic $Y_{km}(\theta, \varphi)$ in coordinates θ and φ can be represented as a linear combination of the spherical harmonics $Y_{km}(\xi, \psi)$ of the same order k:

$$Y_{km}(\theta, \varphi) = \sum_{m'=-k}^{k} a^k_{mm'}\left(\pi, \theta', \frac{\pi}{2} - \varphi'\right) Y_{km'}(\xi, \psi), \tag{3.197}$$

where $a^k_{mm'}$ are the matrix elements of the irreducible representation of the rotation group in H_k. These matrix elements are functions of Euler angles which determine the rotation from the local xyz coordinate system to the original XYZ system. These functions are sometimes called generalized spherical harmonics because they are reduced to spherical harmonics when m or m' is equal to zero.

By using (3.197) and (3.191), we find the expression for P in local coordinates

$$P(\alpha, \beta, \xi, \psi, \theta', \varphi') = \sum_{k=0}^{\infty}\sum_{m=-k}^{k}\sum_{m'=-k}^{k} a^k_{mm'}\left(\pi, \theta', \frac{\pi}{2} - \varphi'\right) P_{km}(\alpha, \beta) Y_{km'}(\xi, \psi). \tag{3.198}$$

By substituting (3.198) into (3.196), we find

$$F(\alpha, \beta, \theta', \varphi') = \sum_{k=0}^{\infty}\sum_{m=-k}^{k}\sum_{m'=-k}^{k} a^k_{mm'}\left(\pi, \theta', \frac{\pi}{2} - \varphi'\right) \times \int_0^{2\pi}\int_0^{\pi/2} \cos\xi P_{km}(\alpha\cos\xi, \beta\cos\xi) Y_{km'}(\xi, \psi)\sin\xi d\xi d\psi. \tag{3.199}$$

In (3.199) the integral with respect to ψ is equal to zero if $m' \neq 0$, and it is equal to 2π if $m' = 0$. Thus, we obtain

$$F(\alpha, \beta, \theta', \varphi') = 2\pi \sum_{k=0}^{\infty}\sum_{m=-k}^{k} a^k_{m0}\left(\pi, \theta', \frac{\pi}{2} - \varphi\right) \times \int_0^{2\pi} \cos\xi P_{km}(\alpha\cos\xi, \beta\cos\xi) Y_{k0}(\xi)\sin\xi d\xi. \tag{3.200}$$

It is known from the group theory (see [67]) that

$$a_{m0}^{k}\left(\pi, \theta', \frac{\pi}{2} - \varphi'\right) = (-1)^m \sqrt{\frac{4\pi}{2k+1}}\, Y_{km}(\theta', \varphi'). \tag{3.201}$$

It is also known (see [62]) that

$$Y_{k0}(\xi) = \sqrt{\frac{2k+1}{4\pi}}\, L_k(\cos \xi), \tag{3.202}$$

where L_k are Legendre polynomials.

From (3.200), (3.201) and (3.202), we derive

$$F(\alpha, \beta, \theta', \varphi') = 2\pi \sum_{k=0}^{\infty} \sum_{m=-k}^{k} (-1)^m Y_{km}(\theta', \varphi') \times \int_0^{\pi/2} P_{km}(\alpha \cos \xi, \beta \cos \xi) L_k(\cos \xi) \cos \xi \sin \xi d\xi. \tag{3.203}$$

By comparing (3.203) with (3.192), we find

$$\int_0^{\pi/2} P_{km}(\alpha \cos \xi, \beta \cos \xi) L_k(\cos \xi) \cos \xi \sin \xi d\xi = \frac{(-1)^m}{2\pi} F_{km}(\alpha, \beta). \tag{3.204}$$

These are the final integral equations which relate P_{km} to F_{km}. These equations can be reduced to Volterra integral equations by using the following change of variables

$$x = \alpha \cos \xi, \qquad \lambda = \frac{\beta}{\alpha}. \tag{3.205}$$

Indeed, after simple transformations, we derive

$$\int_0^{\alpha} P_{km}(x, \lambda x) x L_k\left(\frac{x}{\alpha}\right) dx = (-1)^m \frac{\alpha^2}{2\pi} F_{km}(\alpha, \lambda\alpha). \tag{3.206}$$

Explicit solution to the equations (3.206) can be found for the cases $k = 0$ and $k = 1$. If $k = 0$, from (3.206) we obtain

$$\int_0^{\alpha} x P_{00}(x, \lambda x) dx = \frac{\alpha^2}{2\pi} F_{00}(\alpha, \lambda\alpha). \tag{3.207}$$

By differentiating (3.207) with respect to α, we find

$$P_{00}(\alpha, \lambda\alpha) = \frac{1}{2\pi\alpha} \frac{d}{d\alpha} [\alpha^2 F_{00}(\alpha, \lambda\alpha)]. \tag{3.208}$$

As to be expected, this result coincides with the one obtained for the isotropic model (see (3.168)). If $k = 1$, according to (3.206) we have

$$\int_0^{\alpha} x^2 P_{1m}(x, \lambda x) dx = (-1)^m \frac{\alpha^3}{2\pi} F_{1m}(\alpha, \lambda\alpha), \quad (m = \pm 1, 0). \tag{3.209}$$

By differentiating (3.209), we derive

$$P_{1m}(\alpha, \lambda\alpha) = \frac{(-1)^m}{2\pi\alpha^2} \frac{d}{d\alpha}[\alpha^3 F_{1m}(\alpha, \lambda\alpha)], \quad (m = \pm 1, 0). \tag{3.210}$$

Thus, if we are interested only in the first four terms of spherical harmonic expansion for P (this is a first-order approximation for anisotropic media), then the explicit analytical solution for the identification problem is given by formulas (3.208) and (3.210).

For $k > 1$, equations (3.206) can be solved numerically. Discretization procedures can be applied directly to equations (3.206), or these equations can first be reduced by differentiation to the Volterra equation of the second kind

$$P_{km}(\alpha, \lambda\alpha) - \int_0^\alpha P_{km}(x, \lambda x) \frac{x}{\alpha^3} L_k'\left(\frac{x}{\alpha}\right) dx = \frac{(-1)^m}{2\pi\alpha} \frac{d}{d\alpha}[\alpha^2 F_{km}(\alpha, \lambda\alpha)]. \tag{3.211}$$

In both cases we shall end up with simultaneous algebraic equations with triangular matrices which are easy to solve. However, the reduction to the second kind integral equation may pay off as far as computational stability is concerned.

Finally, the integral equations (3.204) can be solved analytically if polynomial approximations for their right-hand sides are employed. These analytical solutions can be found in exactly the same manner as for 2-D identification problems discussed before.

Up to this point, in order to solve the identification problems, we have used scalar hysteresis data (3.169) and (3.189) measured for unidirectional variations of input $\vec{u}(t)$. However, for anisotropic media these input variations result in vectorial data which for 2-D and 3-D problems can be represented in the following forms, respectively

$$\vec{F}(\alpha, \beta, \varphi) = \tfrac{1}{2}(\vec{f}_{\alpha\varphi} - \vec{f}_{\alpha\beta\varphi}), \tag{3.212}$$

$$\vec{F}(\alpha, \beta, \theta, \varphi) = \tfrac{1}{2}(\vec{f}_{\alpha\theta\varphi} - \vec{f}_{\alpha\beta\theta\varphi}). \tag{3.213}$$

The above vectorial data account for the output components which are orthogonal to the directions of input variations. There is a natural desire to utilize these vectorial data in the identification of vector Preisach models. This can be achieved by generalizing somewhat the models themselves. The essence of generalization is in employing vectorial functions $\vec{\nu}$. This leads to the following vector Preisach model:

$$\vec{f}(t) = \oint\!\!\oint_{|\vec{r}|=1} \left(\iint_{\alpha \geq \beta} \vec{\nu}(\alpha, \beta, \vec{r}) \hat{\gamma}_{\alpha\beta}(\vec{r} \cdot \vec{u}(t)) d\alpha d\beta \right) ds_r. \tag{3.214}$$

By repeating the same reasoning as in the proof of the theorem from Section 2 of this chapter, we can show that by means of redefining $\vec{\nu}$ the integration over a unit sphere in (3.214) can be reduced to the integration over a unit semisphere

$$\vec{f}(t) = \iint_{C^+} \left(\iint_{\alpha \geq \beta} \vec{v}(\alpha, \beta, \vec{r}) \hat{\gamma}_{\alpha\beta}(\vec{r} \cdot \vec{u}(t)) d\alpha d\beta \right) ds_r. \tag{3.215}$$

It can also be shown that the redefined function $\vec{v}$ has the symmetry property

$$\vec{v}(\alpha, \beta, \vec{r}) = -\vec{v}(-\beta, -\alpha, -\vec{r}). \tag{3.216}$$

By using spherical coordinates, the generalized model (3.215) can be represented as

$$\vec{f}(t) = \int_0^{2\pi} \int_0^{\pi/2} \left(\iint_{\alpha \geq \beta} \vec{v}(\alpha, \beta, \theta, \varphi) \hat{\gamma}_{\alpha\beta} u_{\theta\varphi}(t) d\alpha d\beta \right) \sin\theta d\theta d\varphi. \tag{3.217}$$

Similarly, the 2-D model can be expressed in the form

$$\vec{f}(t) = \int_{-\pi/2}^{\pi/2} \left(\iint_{\alpha \geq \beta} \vec{v}(\alpha, \beta, \varphi) \hat{\gamma}_{\alpha\beta} u_{\varphi}(t) d\alpha d\beta \right) d\varphi. \tag{3.218}$$

To solve the identification problems for the models (3.217) and (3.218), we as before introduce the auxiliary functions

$$\vec{P}(\alpha, \beta, \theta, \varphi) = \iint_{T(\alpha, \beta)} \vec{v}(\alpha', \beta', \theta, \varphi) d\alpha' d\beta', \tag{3.219}$$

$$\vec{P}(\alpha, \beta, \varphi) = \iint_{T(\alpha, \beta)} \vec{v}(\alpha', \beta', \varphi) d\alpha' d\beta'. \tag{3.220}$$

These functions can be related to the experimental data (3.213) and (3.212) by the following equations, respectively

$$\int_0^{2\pi} \int_0^{\pi/2} \vec{P}(\alpha \cos\xi, \beta \cos\xi, \xi, \psi, \theta', \varphi') \sin\xi d\xi d\psi = \vec{F}(\alpha, \beta, \theta', \varphi') \tag{3.221}$$

$$\int_{-\pi/2}^{\pi/2} \vec{P}(\alpha \cos\varphi, \beta \cos\psi, \varphi' + \psi) d\psi = \vec{F}(\alpha, \beta, \varphi'). \tag{3.222}$$

By employing sphereical harmonic expansions in the 3-D case and Fourier expansions in the 2-D case, we derive as before the integral equations

$$\int_0^{\pi/2} \vec{P}_{km}(\alpha \cos\psi, \beta \cos\psi) L_k(\cos\xi) \sin\xi d\xi = \frac{(-1)^m}{2\pi} \vec{F}_{km}(\alpha, \beta), \tag{3.223}$$

$$\int_{-\pi/2}^{\pi/2} \cos n\psi \vec{P}_n(\alpha \cos\psi, \beta \cos\psi) d\psi = \vec{F}_n(\alpha, \beta). \tag{3.224}$$

By using the change of variables (3.182) and (3.205), the above integral equations can be reduced to the following forms, respectively:

$$\int_0^\alpha \vec{P}_{km}(x, \lambda x) L_k\left(\frac{x}{\alpha}\right) dx = (-1)^m \frac{\alpha}{2\pi} \vec{F}_{km}(\alpha, \lambda\alpha), \tag{3.225}$$

$$\int_0^\alpha \vec{P}_n(x, \lambda x) \frac{T_n\left(\dfrac{x}{\alpha}\right)}{\sqrt{\alpha^2 - x^2}} dx = \frac{1}{2^n} \vec{F}_n(\alpha, \lambda\alpha). \tag{3.226}$$

When k and n are equal to one, the following explicit solutions of equations (3.225) and (3.226) can be derived:

$$\vec{P}_{1m}(\alpha, \lambda\alpha) = \frac{(-1)^m}{2\pi\alpha} \frac{d}{d\alpha} [\alpha^2 \vec{F}_{1m}(\alpha, \lambda\alpha)], \qquad (m = 0, \pm 1), \tag{3.227}$$

$$\vec{P}_1(\alpha, \lambda\alpha) = \frac{1}{d\alpha} \int_0^\alpha \frac{s^2 \vec{F}_1(s, \lambda s)}{\sqrt{\alpha^2 - s^2}} ds. \tag{3.228}$$

For other values of k and n, polynomial expansion techniques or numerical techniques can be employed for the solution of the above integral equations.

For isotropic media, the data (3.212) and (3.213) are reduced to (3.131). In this case it can be shown that $\vec{v}(\alpha, \beta) = \vec{e}_\varphi v(\alpha, \beta)$ and thus the generalized model (3.218) is reduced to the model (3.67). The proof of the above statement is left to the reader as a useful exercise.

6 Dynamic Vector Preisach Models of Hysteresis

The vector Preisach models of the hysteresis which have so far been discussed are static in nature; they do not account for dynamic properties of vector hysteresis nonlinearities. The purpose of this section is to develop dynamic vector Preisach models of hysteresis. We shall discuss only isotropic dynamic vector models of hysteresis. We begin with 2-D dynamic models; a straightforward extension to three dimensions will then follow.

The main idea of the design of the dynamic vector hysteresis models is to introduce the dependence of the functions v for scalar Preisach models for all directions φ on the speed of variations of output projections df_φ/dt along these directions. This can mathematically be expressed as

$$\vec{f}(t) = \int_{-\pi/2}^{\pi/2} \vec{e}_\varphi \left(\iint\limits_{\alpha \geq \beta} v\left(\alpha, \beta, \frac{df_\varphi}{dt}\right) \hat{\gamma}_{\alpha\beta} u_\varphi(t) \, d\alpha \, d\beta \right) d\varphi. \tag{3.229}$$

The direct utilization of the above model is associated with some untractable difficulties which have been discussed in Section 4 of the previous chapter. These difficulties can be circumvented by using the power series expansion of the v-function with respect to df_φ/dt:

$$v\left(\alpha, \beta, \frac{df_\varphi}{dt}\right) = v_0(\alpha, \beta) + \frac{df_\varphi}{dt} v_1(\alpha, \beta) + \cdots. \tag{3.230}$$

By retaining only the two first terms of the above expansion, we arrive at the following dynamic model:

$$\vec{f}(t) = \vec{f}_0(t) + \int_{-\pi/2}^{\pi/2} \vec{e}_\varphi \frac{df_\varphi}{dt} \left(\iint_{\alpha \geq \beta} v_1(\alpha, \beta) \hat{\gamma}_{\alpha\beta} u_\varphi(t) d\alpha d\beta \right) d\varphi, \tag{3.231}$$

where

$$\vec{f}_0(t) = \int_{-\pi/2}^{\pi/2} \vec{e}_\varphi \left(\iint_{\alpha \geq \beta} v_0(\alpha, \beta) \hat{\gamma}_{\alpha\beta} u_\varphi(t) d\alpha d\beta \right) d\varphi. \tag{3.232}$$

It is clear that in the case of very slow input variations the second term in the right-hand side of (3.231) becomes negligible. Thus, $\vec{f}_0(t)$ can be construed as a static component. This means that the function $v_0(\alpha, \beta)$ should coincide with the v-function of the static model (3.67). In other words, the function $v_0(\alpha, \beta)$ can be determined by matching static first-order transition curves measured for unidirectional variations of the input $\vec{u}(t)$.

We next represent the model (3.231) in cartesian coordinates. To this end, we shall use the following expressions:

$$\vec{e}_\varphi = \vec{e}_x \cos \varphi + \vec{e}_y \sin \varphi, \tag{3.233}$$

$$\frac{df\varphi}{\partial t} = \vec{e}_\varphi \cdot \frac{d\vec{f}}{dt} = \cos \varphi \frac{df_x}{dt} + \sin \varphi \frac{df_y}{dt}. \tag{3.234}$$

By substituting (3.233) and (3.234) into (3.231), after simple transformations we arrive at the following form of the model (3.231):

$$\hat{A} \frac{d\vec{f}}{dt} = \vec{f}(t) - \vec{f}_0(t), \tag{3.235}$$

where the matrix $\hat{A}$ is given by

$$\hat{A} = \begin{pmatrix} \hat{a}_{xx}(\vec{u}(t)) & \hat{a}_{xy}(\vec{u}(t)) \\ \hat{a}_{yx}(\vec{u}(t)) & \hat{a}_{yy}(\vec{u}(t)) \end{pmatrix}, \tag{3.236}$$

and the matrix elements are specified by the expressions:

$$\hat{a}_{xx}(\vec{u}(t)) = \int_{-\pi/2}^{\pi/2} \cos^2 \varphi \left(\iint_{\alpha \geq \beta} v_1(\alpha, \beta) \hat{\gamma}_{\alpha\beta} u_\varphi(t) d\alpha d\beta \right) d\varphi, \tag{3.237}$$

$$\hat{a}_{yy}(\vec{u}(t)) = \int_{-\pi/2}^{\pi/2} \sin^2 \varphi \left(\iint_{\alpha \geq \beta} v_1(\alpha, \beta) \hat{\gamma}_{\alpha\beta} u_\varphi(t) d\alpha d\beta \right) d\varphi, \tag{3.238}$$

$$\hat{a}_{xy}(\vec{u}(t)) = \hat{a}_{yx}(\vec{u}(t)) = \int_{-\pi/2}^{\pi/2} \cos \varphi \sin \varphi \left(\iint_{\alpha \geq \beta} v_1(\alpha, \beta) \hat{\gamma}_{\alpha\beta} u_\varphi(t) d\alpha d\beta \right) d\varphi. \tag{3.239}$$

Thus, the dynamic model (3.231) can be interpreted as a set of two coupled ordinary differential equations (3.235) with hysteretic coefficients (3.237)–(3.239). The expression (3.235) also suggests that the instant speed of output variations is directly proportional to the difference between instant and static output values. The last fact is transparent from the physical point of view.

We next trun to the identification problem of determining the function $\nu_1(\alpha, \beta)$ by fitting the model (3.235) to some experimental data. The following experiments are used to solve this problem. We restrict $\vec{u}(t)$ to vary along the direction $\varphi = 0$ which means that $\vec{u}(t) = \vec{e}_x u(t)$. First, we assume that $u(t)$ is made "infinitely negative" and then it is monotonically increased until it reaches some value α at $t = t_0$. Afterwards, the input is kept constant. As the input is being kept constant, the output relaxes from its value f_α at $t = t_0$ to its static value $f_{0\alpha}$. Due to the symmetry we have

$$\vec{f}_0(t) = \vec{e}_x f_0(t), \qquad \vec{f}(t) = \vec{e}_x f(t). \tag{3.240}$$

Thus, according to the model (3.234)–(3.239), the above relaxation process is described by the differential equation

$$-\tau_\alpha \frac{df}{dt} = f(t) - f_{0\alpha}, \tag{3.241}$$

where

$$\tau_\alpha = \int_{-\pi/2}^{\pi/2} \cos^2 \varphi \left(\iint_{S^-_{\alpha,\varphi}} \nu_1(\alpha', \beta')\,d\alpha'\,d\beta' - \iint_{S^+_{\alpha,\varphi}} \nu_1(\alpha', \beta')\,d\alpha'\,d\beta' \right) d\varphi, \tag{3.242}$$

and geometrical configurations of $S^+_{\alpha,\varphi}$ and $S^-_{\alpha,\varphi}$ are the same as in Fig. 3.28. The solution to the equation (3.241) is given by

$$f(t) = (f_\alpha - f_{0\alpha})e^{-t/\tau_\alpha} + f_{0\alpha}. \tag{3.243}$$

Thus, τ_α has the meaning of relaxation time and can be experimentally measured.

Next the input $u(t)$ is made again "infinitely negative." Then it is monotonically increased until it reaches the value α. Afterwards, the input is monotonically decreased until it reaches some value β at time $t = t'_0$ and it is kept constant for $t > t'_0$. As the input is being kept constant, the output relaxes from its value $f_{\alpha\beta}$ at $t = t'_0$ to its static value $f_{0\alpha\beta}$. Due to the symmetry, expressions (3.240) hold and the model (3.235)–(3.239) yields the following differential equation for the above relaxation process:

$$-\tau_{\alpha\beta} \frac{df}{dt} = f(t) - f_{0\alpha\beta}, \tag{3.244}$$

where

$$\tau_{\alpha\beta} = \int_{-\pi/2}^{\pi/2} \cos^2 \varphi \left(\iint_{S^-_{\alpha,\beta,\varphi}} \nu_1(\alpha', \beta')\,d\alpha'\,d\beta' - \iint_{S^+_{\alpha,\beta,\varphi}} \nu_1(\alpha', \beta')\,d\alpha'\,d\beta' \right) d\varphi, \tag{3.245}$$

and geometry of regions $S^+_{\alpha,\beta,\varphi}$ and $S^-_{\alpha,\beta,\varphi}$ is the same as in Fig. 3.29. By solving (3.244), we find

$$f(t) = (f_{\alpha\beta} - f_{0\alpha\beta})e^{-t/\tau_{\alpha\beta}} + f_{0\alpha\beta}. \tag{3.246}$$

Thus, $\tau_{\alpha\beta}$ has the meaning of relaxation time and can be experimentally measured. We next show that by knowing relaxation times τ_α and $\tau_{\alpha\beta}$ for all possible α and β we can determine the function $\nu_1(\alpha, \beta)$. To this end, we introduce the functions

$$q(\alpha, \beta) = \frac{1}{2}(\tau_\alpha - \tau_{\alpha\beta}), \tag{3.247}$$

$$P_1(\alpha, \beta) = \iint\limits_{T(\alpha,\beta)} \nu_1(\alpha', \beta')d\alpha' d\beta', \tag{3.248}$$

where $T(\alpha, \beta)$ is a triangle shown in Fig. 3.10.

It is clear as before that P_1 and ν_1 are related by the formula

$$\nu_1(\alpha, \beta) = -\frac{\partial^2 P_1(\alpha, \beta)}{\partial\alpha\partial\beta}. \tag{3.249}$$

Thus, if the function $P_1(\alpha, \beta)$ is found, then the function $\nu_1(\alpha, \beta)$ can be retrieved. However, from the computational point of view, it is more convenient to deal with the function $P_1(\alpha, \beta)$ rather than with $\nu_1(\alpha, \beta)$. This is because the double integral with respect to α and β in expressions (3.237)–(3.239) can be explicitly expressed in terms of $P_1(\alpha, \beta)$ by using formulas similar to (3.71). Another advantage of using $P_1(\alpha, \beta)$ is that this function can be directly related to the experimental data (3.247). Indeed, by using the expression (3.242), (3.245), (3.247), and (3.248) and Fig. 3.28 and 3.29, we derive

$$\int_{-\pi/2}^{\pi/2} \cos^2 \varphi P_1(\alpha \cos \varphi, \beta \cos \varphi)d\varphi = q(\alpha, \beta). \tag{3.250}$$

The expression (3.250) is the integral equation which relates the function $P_1(\alpha, \beta)$ to experimental data $q(\alpha, \beta)$. This equation is similar to the integral equation (3.132) and, consequently, the same techniques can be used for the solution of the equation (3.250) as for the equation (3.132). Namely, by using the change of variables

$$x = \alpha \cos \varphi, \qquad \lambda = \beta/\alpha, \tag{3.251}$$

the equation (3.250) can be reduced to the following Abel-type integral equation

$$\int_0^\alpha P_1(x, \lambda x)\frac{x^2 dx}{\sqrt{\alpha^2 - x^2}} = \frac{\alpha^2}{2} q(\alpha, \lambda\alpha). \tag{3.252}$$

By using the technique discussed in Section 4 of this chapter, the following closed-form solution of the above equation can be found:

$$P_1(\alpha, \lambda\alpha) = \frac{1}{\pi\alpha^2}\frac{d}{d\alpha}\int_0^\alpha \frac{s^3 q(s, \lambda s)}{\sqrt{\alpha^2 - s^2}}dx. \tag{3.253}$$

We next turn to the discussion of three-dimensional dynamic vector Preisach models of hysteresis. Similar to (3.231), these models can be represented in the following mathematical form

$$\vec{f}(t) = \vec{f}_0(t) + \int_0^{2\pi} \int_0^{\pi/2} \vec{e}_{\theta\varphi} \frac{df_{\theta\varphi}}{dt} \left(\iint_{\alpha \geq \beta} v_1(\alpha, \beta) \hat{\gamma}_{\alpha\beta} u_{\theta\varphi}(t) d\alpha d\beta \right) \sin\theta d\theta d\varphi, \tag{3.254}$$

where $\vec{f}_0(t)$ represents a static component of hysteresis nonlinearities defined by the expression

$$\vec{f}_0(t) = \int_0^{2\pi} \int_0^{\pi/2} \vec{e}_{\theta\varphi} \left(\iint_{\alpha \geq \beta} v_0(\alpha, \beta) \hat{\gamma}_{\alpha\beta} u_{\theta\varphi}(t) d\alpha d\beta \right) \sin\theta d\theta d\varphi, \tag{3.255}$$

By using cartesian coordinates, we have

$$\vec{e}_{\theta\varphi} = \vec{e}_x \cos\varphi \sin\theta + \vec{e}_y \sin\varphi \sin\theta + \vec{e}_z \cos\theta, \tag{3.256}$$

$$\frac{df_{\theta\varphi}}{dt} = \cos\varphi \sin\theta \frac{df_x}{dt} + \sin\varphi \sin\theta \frac{df_y}{dt} + \cos\theta \frac{df_z}{dt}. \tag{3.257}$$

By substituting (3.256) and (3.257) into (3.254), after simple transformations we find

$$\hat{A} \frac{d\vec{f}}{dt} = \vec{f}(t) - \vec{f}_0(t), \tag{3.258}$$

where the matrix $\hat{A}$ and its elements are given by

$$\hat{A} = \begin{pmatrix} \hat{a}_{xx}(\vec{u}(t)) & \hat{a}_{xy}(\vec{u}(t)) & \hat{a}_{xz}(\vec{u}(t)) \\ \hat{a}_{yx}(\vec{u}(t)) & \hat{a}_{yy}(\vec{u}(t)) & \hat{a}_{yz}(\vec{u}(t)) \\ \hat{a}_{zx}(\vec{u}(t)) & \hat{a}_{zy}(\vec{u}(t)) & \hat{a}_{zz}(\vec{u}(t)) \end{pmatrix}, \tag{3.259}$$

$$\hat{a}_{xx}(\vec{u}(t)) = \int_0^{2\pi} \int_0^{\pi/2} \cos^2\varphi \sin^3\theta \left(\iint_{\alpha \geq \beta} v_1(\alpha, \beta) \hat{\gamma}_{\alpha\beta} u_{\theta\varphi}(t) d\alpha d\beta \right) d\theta d\varphi, \tag{3.260}$$

$$\hat{a}_{yy}(\vec{u}(t)) = \int_0^{2\pi} \int_0^{\pi/2} \sin^2\varphi \sin^3\theta \left(\iint_{\alpha \geq \beta} v_1(\alpha, \beta) \hat{\gamma}_{\alpha\beta} u_{\theta\varphi}(t) d\alpha d\beta \right) d\theta d\varphi, \tag{3.261}$$

$$\hat{a}_{zz}(\vec{u}(t)) = \int_0^{2\pi} \int_0^{\pi/2} \cos^2\theta \sin\theta \left(\iint_{\alpha \geq \beta} v_1(\alpha, \beta) \hat{\gamma}_{\alpha\beta} u_{\theta\varphi}(t) d\alpha d\beta \right) d\theta d\varphi, \tag{3.262}$$

$$\begin{aligned} \hat{a}_{xy}(\vec{u}(t)) &= \hat{a}_{yx}(\vec{u}(t)) \\ &= \int_0^{2\pi} \int_0^{\pi/2} \cos\varphi \sin\varphi \sin^3\theta \left(\iint_{\alpha \geq \beta} v_1(\alpha, \beta) \hat{\gamma}_{\alpha\beta} u_{\theta\varphi}(t) d\alpha d\beta \right) d\theta d\varphi, \end{aligned} \tag{3.263}$$

$$\hat{a}_{xz}(\vec{u}(t)) = \hat{a}_{zx}(\vec{u}(t))$$
$$= \int_0^{2\pi} \int_0^{\pi/2} \cos\varphi \cos\theta \sin^2\theta \left(\iint_{\alpha \geq \beta} v_1(\alpha, \beta) \hat{\gamma}_{\alpha\beta} u_{\theta\varphi}(t) d\alpha d\beta \right) d\theta d\varphi, \tag{3.264}$$

$$\hat{a}_{yz}(\vec{u}(t)) = \hat{a}_{zy}(\vec{u}(t))$$
$$= \int_0^{2\pi} \int_0^{\pi/2} \sin\varphi \cos\theta \sin^2\theta \left(\iint_{\alpha \geq \beta} v_1(\alpha, \beta) \hat{\gamma}_{\alpha\beta} u_{\theta\varphi}(t) d\alpha d\beta \right) d\theta d\varphi, \tag{3.265}$$

Thus, the 3-D dynamic vector Preisach model of hysteresis (3.254) can be interpreted as a set of coupled ordinary differential equations (3.258) with hysteretic coefficients (3.260)–(3.265).

We next proceed to the discussion of the identification problem for the model (3.258)–(3.265). The solution to this problem is very similar to that for the model (3.235)–(3.239) except that the final form of the solution is much simpler in three dimensions than in two dimensions. The experimental data used for the identification of the model (3.258)–(3.265) is measured when the input is restricted to vary along the axis z, that is, when $\vec{u}(t) = \vec{e}_z u(t)$. As before, two types of relaxation processes are considered. The relaxation processes of first type occur when the input is made "infinitely negative" and then monotonically increased to some value α and kept constant thereafter. The relaxation processes of the second type occur when, starting from the state of negative saturation along the axis z, the input is first monotonically increased to some value α, then monotonically decreased to some value β and kept constant afterwards. The relaxation times τ_α and $\tau_{\alpha\beta}$ of the above processes can be measured and used for computing the function $q(\alpha, \beta)$ defined by (3.247). On the grounds of symmetry, for both types of relaxation processes we have

$$\vec{f}_0 = \vec{e}_z f_0(t), \qquad \vec{f}(t) = \vec{e}_z f(t). \tag{3.266}$$

This results in the reduction of the model (3.258)–(3.265) to the equation

$$\hat{a}_{zz}(\vec{u}(t)) \frac{df}{dt} = f(t) - f_0(t). \tag{3.267}$$

Now, by using the same line of reasoning as in the derivation of the equation (3.250), we can show that the function $P_1(\alpha, \beta)$ defined by (3.248) is related to the experimental data $q(\alpha, \beta)$ by the integral equation

$$\int_0^{\pi/2} P_1(\alpha\cos\theta, \beta\cos\theta)\cos^2\theta \sin\theta d\theta = \frac{q(\alpha, \beta)}{2\pi}. \tag{3.268}$$

By using the change of variables

$$x = \alpha\cos\theta, \qquad \lambda = \frac{\beta}{\alpha}, \tag{3.269}$$

the integral equation (3.268) is reduced to the form

$$\int_0^\alpha x^2 P_1(x, \lambda x)dx = \frac{\alpha^3}{2\pi} q(\alpha, \lambda\alpha), \tag{3.270}$$

from which we derived the final expression

$$P_1(x, \lambda x) = \frac{1}{2\pi\alpha^2} \frac{d}{d\alpha} [\alpha^3 q(\alpha, \lambda\alpha)]. \tag{3.271}$$

7 Generalized Vector Preisach Models of Hysteresis

In the construction of vector Preisach models we have so far used classical scalar Preisach models continuously distributed along all possible directions. A natural way to further generalize the above vector Preisach models is to use generalized scalar Preisach models instead of classical ones. Such generalized scalar Preisach models have been discussed in Chapter II and any of them can be employed for the above purpose. To be specific, we shall use the nonlinear (input-dependent) scalar Preisach models discussed in Section 2 of the preivous chapter. Analysis of generalized vector Preisach models is very similar to that for the "classical" vector Preisach models. For this reason, our discussion will be concise and will mainly be centered around the description of final results, while filling in the details will be left to the reader.

We begin with 2-D isotropic models which can be represented in the mathematical form

$$\vec{f}(t) = \int_{-\pi/2}^{\pi/2} \vec{e}_\varphi \left(\iint_{R_{u_\varphi}(t)} \nu(\alpha, \beta)\, u_\varphi(t)) \hat{\gamma}_{\alpha\beta} u_\varphi(t) d\alpha d\beta \right) d\varphi + \xi(u(t))\vec{u}(t). \tag{3.272}$$

The numerical implementation of the above model is substantially facilitated by use of the function

$$P(\alpha, \beta, u) = \iint_{R_{\alpha\beta u}} \nu(\alpha', \beta', u) d\alpha' d\beta', \tag{3.273}$$

where $R_{\alpha\beta u}$ is a rectangle shown in Fig. 2.12.

By employing the above function, we can find explicit expressions for the double integral with respect to α and β in (3.272). These expressions are similar to formula (2.66). In addition, the function $P(\alpha, \beta, u)$ can be directly used for the identification of the model (3.272). First- and second-order transition curves measured along any fixed direction (for instance, along the direction $\varphi = 0$) will be utilized for the solution of the identification problem. By using these curves, the following function can be constructed:

$$F(\alpha, \beta, u) = \frac{1}{2}(f_{\alpha u} - f_{\alpha\beta u}), \tag{3.274}$$

where $f_{\alpha u}$ and $f_{\alpha\beta u}$ have the same meaning as in Section 2 of the previous chapter.

It can be shown that the function $P(\alpha, \beta, u)$ is related to the experimental data (3.274) by the integral equation

$$\int_{-\pi/2}^{\pi/2} \cos\varphi P(\alpha\cos\varphi, \beta\cos\varphi, u\cos\varphi)d\varphi = F(\alpha, \beta, u) \tag{3.275}$$

This equation is very similar to the integral equation (3.132). Thus, the same techniques can be employed for the solution of equation (3.275) as for the solution of equation (3.132). For instance, an approximate polynomial solution of the above equation can be found. The finding of this solution is based on the fact that monomials $\alpha^k\beta^s u^m$ are eigenfunctions of the operator $\hat{A}$:

$$\hat{A}P = \int_{-\pi/2}^{\pi/2} \cos\varphi P(\alpha\cos\varphi, \beta\cos\varphi, u\cos\varphi)d\varphi. \tag{3.276}$$

In other words, we have

$$\hat{A}\alpha^k\beta^s u^m = \lambda_{k+s+m}\alpha^k\beta^s u^m, \tag{3.277}$$

where

$$\lambda_{k+s+m} = \int_{-\pi/2}^{\pi/2} (\cos\varphi)^{k+s+m+1}d\varphi. \tag{3.278}$$

Now, by expanding the right-hand side of the integral equation (3.275) into the series of Chebyshev polynomials

$$F(\alpha, \beta, u) = \sum_{\ell,q,m} a_{\ell qm}T_\ell(\alpha)T_q(\beta)T_m(u), \tag{3.279}$$

we can represent the solution of the above equation in the form

$$P(\alpha, \beta, u) = \sum_{\ell,q,m} a_{\ell qm}t_{\ell qm}(\alpha, \beta, u), \tag{3.280}$$

where the polynomials $t_{\ell qm}(\alpha, \beta, u)$ are mapped by operator $\hat{A}$ into polynomials $T_\ell(\alpha)T_q(\beta)T_m(u)$. Polynominals $t_{\ell qm}(\alpha, \beta, u)$ can be determined by using expressions (3.277) and (3.278). As far as expansion coefficients $a_{\ell qm}$ are concerned, they can be computed by using the formula similar to (3.146):

$$a_{\ell qm} = \frac{8}{\pi^3}\int_0^\pi\int_0^\pi\int_0^\pi F(\cos\theta, \cos\theta', \cos\theta'')\cos\ell\theta\cos q\theta'\cos m\theta''d\theta d\theta' d\theta''. \tag{3.281}$$

It is also possible to find a closed-form solution to the equation (3.275). To this end, the following change of variables is used:

$$x = \alpha\cos\varphi, \qquad \lambda = \beta/\alpha, \qquad \chi = u/\alpha, \tag{3.282}$$

and the above equation is reduced to the Abel type integral equation

$$\int_0^\pi \frac{x}{\sqrt{\alpha^2 - x^2}}P(x, \lambda x, \chi x)dx = \frac{\alpha}{2}F(\alpha, \lambda\alpha, \chi\alpha). \tag{3.283}$$

Using the same reasoning as in Section 4 of this chapter, we obtain the following solution of equation (3.283):

$$P(\alpha, \lambda\alpha, \chi\alpha) = \frac{1}{\pi\alpha}\frac{d}{d\alpha}\int_0^\alpha \frac{s^2 F(s, \lambda s, \chi s)}{\sqrt{\alpha^2 - s^2}}\,ds. \tag{3.284}$$

As is seen from the above discussion, function $P(\alpha, \beta, u)$ (and consequently function $\nu(\alpha, \beta, u)$ can be determined by matching the increments (3.274) between the first-order and second-order transition curves. The function $\xi(u)$ in (3.272) can be found by matching an ascending branch of a major loop. This leads to the expression

$$\xi(u) = \frac{f_u^+ + f_u^-}{2u} \tag{3.285}$$

We next proceed to the discussion of 3-D generalized vector Preisach models of hysteresis. These models can be represented as

$$\vec{f}(t) = \int_0^{2\pi}\int_0^{\pi} \vec{e}_{\theta\varphi}\left(\iint_{R_{u_{\theta\varphi}}(t)} \nu(\alpha, \beta, u_{\theta\varphi}(t))\hat{\gamma}_{\alpha\beta}u_{\theta\varphi}(t)\,d\alpha\,d\beta\right)\sin\theta\,d\theta\,d\varphi + \xi(u(t))\vec{u}(t). \tag{3.286}$$

By introducing function P defined by (3.273) and by using experimental data (3.274) the identification problem for the model (3.286) can be reduced to the integral equation

$$\int_0^{\pi/2} \cos\theta \sin\theta P(\alpha\cos\theta, \beta\cos\theta, u\cos\theta)\,d\theta = \frac{F(\alpha, \beta, u)}{2\pi}. \tag{3.287}$$

The explicit solution of the above equation can be obtained by using the change of variables

$$x = \alpha\cos\theta, \qquad \lambda = \beta/\alpha, \qquad \chi = u/\alpha, \tag{3.288}$$

and by reducing (3.287) to the integral equation

$$\int_0^\alpha xP(x, \lambda x, \chi x)\,dx = \frac{\alpha^2}{2\pi}F(\alpha, \lambda\alpha, \chi\alpha). \tag{3.289}$$

Differentiation of (3.288) yield

$$P(\alpha, \lambda\alpha, \chi\alpha) = \frac{1}{2\pi\alpha}\frac{d}{d\alpha}[\alpha^2 F(\alpha, \lambda\alpha, \chi\alpha)]. \tag{3.290}$$

We next turn to generalized dynamic vector Preisach models of hysteresis. For the sake of notational simplicity, we consider only 2-D models; extensions to 3-D models are straightforward. The 2-D generalized dynamic Preisach models can be defined as

$$\vec{f}(t) = \vec{f}_0(t) + \int_{-\pi/2}^{\pi/2} \vec{e}_\varphi \frac{df\varphi}{dt}\left(\iint_{R_{u_\varphi}(t)} \nu_1(\alpha, \beta, u_\varphi(t))\hat{\gamma}_{\alpha\beta}u_\varphi(t)\,d\alpha\,d\beta\right)d\varphi, \tag{3.291}$$

where $\vec{f}_0$ is a static component of hysteresis nonlinearity which coincides with (3.272).

By using cartesian coordinates, the model (3.291) can be reduced to the ordinary differential equations

$$\hat{A}\frac{d\vec{f}}{dt} = \vec{f}(t) - \vec{f}_0(t), \tag{3.292}$$

where the matrix $\hat{A}$ has the following entries:

$$\hat{a}_{xx}(\vec{u}(t)) = \int_{-\pi/2}^{\pi/2} \cos^2\varphi\left(\iint_{R_{u_\varphi}(t)} \nu_1(\alpha, \beta, u_\varphi(t))\hat{\gamma}_{\alpha\beta}u_\varphi(t)d\alpha d\beta\right)d\varphi, \tag{3.293}$$

$$\hat{a}_{yy}(\vec{u}(t)) = \int_{-\pi/2}^{\pi/2} \sin^2\varphi\left(\iint_{R_{u_\varphi}(t)} \nu_1(\alpha, \beta, u_\varphi(t))\hat{\gamma}_{\alpha\beta}u_\varphi(t)d\alpha d\beta\right)d\varphi, \tag{3.294}$$

$$\hat{a}_{xy}(\vec{u}(t)) = \hat{a}_{yx}(\vec{u}(t)) = \int_{-\pi/2}^{\pi/2} \cos\varphi\sin\varphi\left(\iint_{R_{u_\varphi}(t)} \nu_1(\alpha, \beta, u_\varphi(t))\hat{\gamma}_{\alpha\beta}u_\varphi(t)d\alpha d\beta\right)d\varphi. \tag{3.295}$$

To solve the identification problem for the model (3.292)–(3.295), we restrict input to vary along the axis x. Then, on the grounds of symmetry, it is easy to conclude that

$$\vec{f}(t) = \vec{e}_x f(t), \qquad \vec{f}_0(t) = \vec{e}_x f_0(t). \tag{3.296}$$

This results in the reduction of the model (3.292)–(3.295) to the equation

$$\hat{a}_{xx}(u(t))\frac{df}{dt} = f(t) - f_0(t). \tag{3.297}$$

We next introduce the function $P_1(\alpha, \beta, u)$:

$$P_1(\alpha, \beta, u) = \iint_{R_{\alpha\beta u}} \nu_1(\alpha', \beta', u)d\alpha' d\beta', \tag{3.298}$$

and experimental data

$$Q(\alpha, \beta, u) = \frac{1}{2}(\tau_{\alpha u} - \tau_{\alpha\beta u}), \tag{3.299}$$

where $\tau_{\alpha u}$ and $\tau_{\alpha\beta u}$ are relaxation times for the processes which are described in detail in Section 4 of the previous chapter. By using the expressions (3.298)–(3.299) and the same line of reasoning as in the previous section, it can be shown that the function $P_1(\alpha, \beta, u)$ is related to the experimental data (3.299) by the integral equation

$$\int_{-\pi/2}^{\pi/2} \cos^2\varphi P_1(\alpha\cos\varphi, \beta\cos\varphi, u\cos\varphi)d\varphi = Q(\alpha, \beta, u). \tag{3.300}$$

By using the change of variables (3.282), the above equation can be reduced to the Abel-type integral equation

$$\int_0^\alpha \frac{x^2}{\sqrt{\alpha^2 - x^2}} P_1(x, \lambda x, \chi x)dx = \frac{\alpha^2}{2} Q(\alpha, \lambda\alpha, \chi\alpha), \tag{3.301}$$

whose solution is given by

$$P_1(\alpha, \lambda\alpha, \chi\alpha) = \frac{1}{\pi\alpha^2} \frac{d}{d\alpha} \int_0^\alpha \frac{s^3 Q(s, \lambda s, \chi s)}{\sqrt{\alpha^2 - s^2}} ds. \tag{3.302}$$

Finally, we shall discuss generalized anisotropic vector Preisach models. Again, for the sake of notational simplicity, we consider only 2-D models. These models can be defined as

$$\vec{f}(t) = \int_{-\pi/2}^{\pi/2} \left(\iint_{R_{u_\varphi}(t)} \vec{v}(\alpha, \beta, \varphi, u_\varphi(t)) \hat{\gamma}_{\alpha\beta} u_\varphi(t) d\alpha d\beta \right) d\varphi + \vec{\xi}(\vec{u}(t)). \tag{3.303}$$

The experimental data defined by the function

$$\vec{F}(\alpha, \beta, u, \varphi) = \frac{1}{2}(\vec{f}_{\alpha u\varphi} - \vec{f}_{\alpha\beta u\varphi}) \tag{3.304}$$

will be used for the identification of the models (3.303).

As before, we introduced the auxiliary function

$$\vec{P}(\alpha, \beta, \varphi, u) = \iint_{R_{\alpha\beta u}} \vec{v}(\alpha', \beta', \varphi, u) d\alpha' d\beta'. \tag{3.305}$$

We shall employ the following Fourier expansions:

$$\vec{P}(\alpha, \beta, u, \varphi) = \sum_{n=-\infty}^{\infty} \vec{P}_n(\alpha, \beta, u) e^{in\varphi}, \tag{3.306}$$

$$\vec{F}(\alpha, \beta, u, \varphi) = \sum_{n=-\infty}^{\infty} \vec{F}_n(\alpha, \beta, u) e^{in\varphi}. \tag{3.307}$$

By using (3.304)–(3.307), it can be shown that $\vec{P}_n$ are related to $\vec{F}_n$ by the integral equations

$$\int_{-\pi/2}^{\pi/2} \cos n\psi \vec{P}_n(\alpha \cos\psi, \beta \cos\psi, u \cos\psi) d\psi = \vec{F}_n(\alpha, \beta, u). \tag{3.308}$$

These equations can be reduced to the Abel-type integral equations

$$\int_0^\alpha \vec{P}_n(x, \lambda x, \chi x) \frac{T_n\left(\dfrac{x}{\alpha}\right)}{\sqrt{\alpha^2 - x^2}} dx = \frac{1}{2^n} \vec{F}_n(\alpha, \lambda\alpha, \chi\alpha). \tag{3.309}$$

Different numerical techniques can be used for the solution of these equations. In particular case when $n = \pm 1$, the following closed-form solutions can be found:

$$\vec{P}_{\pm 1}(\alpha, \lambda\alpha, \chi\alpha) = \frac{1}{\pi\alpha}\frac{d}{d\alpha}\int_0^\alpha \frac{s^2\vec{F}_{\pm 1}(s, \lambda s, \chi s)}{\sqrt{\alpha^2 - s^2}}ds. \tag{3.310}$$

As far as the function $\vec{\xi}(\vec{u})$ in (3.303) is concerned, the following expression (similar to (3.285)) can be derived:

$$\vec{\xi}(\vec{e}_\varphi u) = \frac{\vec{f}^{+}_{u,\varphi} + \vec{f}^{-}_{u,\varphi}}{2}, \tag{3.311}$$

where the notations in (3.311) are self-explanatory.

We next consider vector Preisach models driven by vector stochastic inputs as models for viscosity. Below we shall only discuss the case of 2-D vector Preisach model (3.65). Extensions to the 3-D case will be striaghtforward and, for this reason, they will be omitted.

Consider a deterministic vector input $\vec{u}(t)$ which at time $t = 0$ assumes some value $\vec{u}$ and remains constant thereafter. We assume that some zero mean vector noise described by a discrete-time random process $\vec{V}_n$ is superimposed on $\vec{u}(t)$. As a result, the Preisach model (3.65) is driven by the process:

$$\vec{v}_n = \vec{u} + \vec{V}_n \tag{3.312}$$

For the output process we have:

$$\vec{f}_n = \int_{-\pi/2}^{\pi/2} \vec{e}_\varphi\left(\iint_{\alpha \geq \beta} \nu(\alpha, \beta, \varphi)\hat{\gamma}_{\alpha\beta}v_{n\varphi}\,d\alpha\,d\beta\right)d\varphi, \tag{3.313}$$

From (3.313), we derive the following expression for the expected value of $\vec{f}_n$:

$$E\{\vec{f}_n\} = \int_{-\pi/2}^{\pi/2} \vec{e}_\varphi\left(\iint_{\alpha \geq \beta} \nu(\alpha, \beta, \varphi)E\{\hat{\gamma}_{\alpha\beta)\beta}v_{n\varphi}\}\,d\alpha\,d\beta\right)d\varphi, \tag{3.314}$$

By modelling the noise by i.i.d. vector process with independent cartesian components and by repeating the same line of reasoning as in Section 6 of Chaper I, we derive:

$$E\{\hat{\gamma}_{\alpha\beta}v_{n\varphi}\} = [\vartheta(\alpha, \beta, \varphi) - \zeta(\alpha, \beta, \varphi)]r^n_{\alpha\beta\varphi} + \zeta(\alpha, \beta, \varphi). \tag{3.315}$$

where

$$\vartheta(\alpha, \beta, \varphi) = \begin{cases} +1, & \text{if } (\alpha, \beta) \in S^+_\varphi(0), \\ -1, & \text{if } (\alpha, \beta) \in S^-_\varphi(0). \end{cases} \tag{3.316}$$

$$\zeta(\alpha, \beta, \varphi) = \frac{P^{-+}_{\alpha\varphi} - P^{+-}_{\beta\varphi}}{P^{-+}_{\alpha\varphi} + P^{+-}_{\beta\varphi}}, \tag{3.317}$$

$$P^{-+}_{\alpha\varphi} = \int_\alpha^\infty \rho_\varphi(x)dx, \tag{3.318}$$

$$P_{\beta\varphi}^{+-} = \int_{-\infty}^{\beta} \rho_\varphi(x)dx, \tag{3.319}$$

$$r_{\alpha\beta\varphi} = \int_{\beta}^{\alpha} \rho_\varphi(x)dx. \tag{3.320}$$

In expressions (3.318)–(3.320), ρ_φ is a probability density function for $v_{n\varphi}$. This function can be expressed in terms of probability density functions for cartesian components of $\vec{v}_n$. Indeed, for $v_{n\varphi}$ we have:

$$v_{n\varphi} = x_n \cos\varphi + y_n \sin\varphi \tag{3.321}$$

where x_n and y_n are independent cartesian components of $\vec{v}_n$ with probability density functions $\rho_x(x)$ and $\rho_y(y)$, respectively. According to the well known result from the probability theory, from (3.321) we derive:

$$\rho_\varphi(v) = \frac{1}{\cos\varphi \sin\varphi} \int_{-\infty}^{+\infty} \rho_x\left(\frac{x}{\cos\varphi}\right) \rho_y\left(\frac{v-x}{\sin\varphi}\right) dx \tag{3.322}$$

By substituting (3.315) into (3.314) and by replacing discrete time n by continuous time t, after simple transformations we obtain:

$$E\{\vec{f}_y\} = E\{\vec{f}_\infty\} + \int_{-\pi/2}^{\pi/2} \vec{e}_\varphi \left(\iint_{\alpha \geq \beta} \chi(\alpha, \beta, \varphi) e^{-\xi(\alpha,\beta,\varphi)t} d\alpha d\beta \right) d\varphi, \tag{3.323}$$

where

$$E\{\vec{f}_\infty\} = \int_{-\pi/2}^{\pi/2} \vec{e}_\varphi \left(\iint_{\alpha \geq \beta} \nu(\alpha, \beta, \varphi)\zeta(\alpha, \beta, \varphi) d\alpha d\beta \right) d\varphi, \tag{3.324}$$

$$\chi(\alpha, \beta, \varphi) = \nu(\alpha, \beta, \varphi)[\vartheta(\alpha, \beta, \varphi) - \zeta, \beta, \varphi)], \tag{3.325}$$

$$\xi(\alpha, \beta, \varphi) = -\frac{1}{\Delta t} \ln r_{\alpha\beta\varphi}. \tag{3.326}$$

By comparing (3.323) with (1.190), it can be concluded that the model (3.323) can be regarded as a vector generalization of the classical thermal activation type models.

For the isotropic model the symmetry property (3.58) holds. It can also be checked that for $\vec{u} = 0$, ρ_φ is an even function if ρ_x and ρ_y are even. By using these facts and the same line of reasoning as Section 6 of Chapter I, it can be proven that:

$$E\{\vec{f}_\infty\} = 0 \tag{3.327}$$

Therefore, for $\vec{u} = 0$ and regardless of the past history, the limiting expected value of output equals to zero.

Summary

In the conclusion of this chapter, a comparison between the vector Preisach models and the classical Stoner-Wohlfarth model suggests itself. It seems that the vector Preisach models have the following advantages.

1. The vector Preisach models are constructed as superpositions of scalar Preisach models. Since the scalar Preisach models describe nonsymmetrical minor loops, the same property is inherited by the vector Preisach models. In this manner, the vector Preisach models take the "particle interactions" into account on the phenomenological level.
2. If the magnetic field is restricted to vary along only one direction, then the vector models are reduced to scalar Preisach models. This reduction is very valuable because the saclar Preisach model has long been regarded as a fairly good model for scalar hysteresis.
3. The vector Preisach models are quite efficient from a computational point of view. This is because explicit (avoiding integration) expressions exist for the outputs of scalar Preisach models in terms of directly measured experimental data.
4. There are well established procedures for the solution of the identification problems for the vector Preisach models. These problems can be reduced to the solution of special integral equations. Efficient analytical techniques for the solution of these integral equations are presented in this chapter.

References

1. Einstein A. Geometry and Experience, in *Sidelights on Reality*, Dover Publications, New York, 1983.
2. Krasnoselskii M., Pokrovskii A. *Systems with Hysteresis*, Nauka, Moscow, 1983.
3. Stoner E.C., Wolhfarth E.P. *Trans. R. Soc. London* 240, p. 599, 1948.
4. Madelung E. *Ann Physik* 17, p. 865, 1905.
5. Everett D.H., Smith F.W. *Trans. Faraday Soc.* 50, p. 187, 1954.
6. Potter R.I., Schmulian R.J. *IEEE Trans. Magnetics.* 7, p. 873, 1971.
7. Chua L., Stromsmoe K. *IEEE Trans. Circuit Theory* 17, p. 564, 1970.
8. Chua L., Bass S. *IEEE Trans. Circuit Theory* 19, p. 36, 1972.
9. Boley C.D., Hodgdon M.L. *IEEE Trans. Magnetics* 25, p. 3922, 1989.
10. Jiles D.C., Thoelke J.B. *IEEE Trans. Magnetics* 25, p. 3928, 1989.
11. Poincaré H. *Science and Hypothesis*, Dover Publications, New York, 1952.
12. Preisach F. *Z. Phys.* 94, p. 277, 1935.
13. Neel L. *Compt. Rend. Acad. Sci. Paris* 246, pp. 2313, 1958.
14. Biorci G., Pescetti D. *Nuovo Cinento* 7, p. 829, 1958.
15. Biorci G., Pescetti D. *J. Phys. Radium* 20, p. 233, 1959.
16. Biorci G., Pescetti D. *J. Appl. Phys.* 37, p. 425, 1966.
17. Brown W.F. Jr. *J. Appl. Phys.* 33, p. 1308, 1962.
18. Bate G. *J. Appl. Phys.* 33, p. 2263, 1962.
19. Woodward J.G., Della Torre E. *J. Appl. Phys.* 31, p. 56, 1960.
20. Della Torre E. *J. Appl. Phys.* 36, p. 518, 1965.
21. Barker J.A., Schreiber D.E., Huth B.G., Everett D.H. *Proc. R. Soc. London* A386, p. 251, 1985.
22. Damlanian A., Visintin A. *Compt. Rend. Acad. Sci. Paris* 297, p. 437, 1983.
23. Everett D.H., Whitton W.I. *Trans. Faraday Soc.* 48, p. 749, 1952.
24. Everett D.H. *Trans. Faraday Soc.* 50, p. 1077, 1954.
25. Everett D.H. *Trans. Faraday Soc.* 51, p. 1551, 1955.
26. Enderby J.A. *Trans. Faraday Soc.* 52, p. 106, 1956.
27. Friedman A. *Foundation of Modern Analysis*, Dover Publications, New York, 1982.
28. Kozjakin V.S., Krasnoselskii M.A., Pokrovskii A.V. *Soviet Math. Dokl.* 13, p. 1305, 1972.
29. Brokate M. *IEEE Trans. Magnetics*, 25, p. 2922, 1989.
30. Brokate M., Visintin A. *J. für Rheine U. Angew. Math* (in press).
31. Brokate M., Friedman A. *SIAM J. Control and Optimization* 27, p. 697, 1989.
32. Visintin A. *Nonlinear Analysis* 9, p. 977, 1984.
33. Visintin A.A. *Ann. Mat. Pura Appl.* 131, p. 203, 1982.

34. Friedman A., Karl-Heinz Hoffmann. *SIAM J. Control and Optimization*, 26, p. 42, 1988.
35. Wang Z.Z., Ong N.P. *Physica B&C* 143B, p. 100, 1986.
36. Hopfield J.J. *Proc. Natl. Acad. Sci., USA* 79, p. 2554, 1982.
37. Del Vecchio R.M. *IEEE Trans. Magnetics* 16, p. 809, 1980.
38. Del Vecchio R.M. *IEEE Trans. Magnetics* 18, p. 275, 1982.
39. Del Vecchio R.M. *J. Appl. Phys.* 53, p. 8281, 1982.
40. Prigogine I. *Introduction to Thermodynamics of Irreversible Processes*, Wiley, New York, 1961.
41. deGroot R., Mazur P. *Non-Equilibrium Thermodynamics*, North-Holland, Amsterdam, 1963.
42. Vincenti W.G., Kruger C.H. Jr. *Introduction to Physical Gas Dynamics*, Wiley, New York, 1965.
43. Cory J.S., McNichols J.L. *J. Appl. Phys.* 58, p. 3282, 1985.
44. Charap S.H. *J. Appl. Phys.* 63, p. 2054, 1988.
45. Wiesen K., Charap S. *IEEE Trans. Magnetics* 24, p. 249, 1988.
46. Wiesen K., Charap S.H. *J. Appl. Phys.* 61, p. 4019, 1987.
47. Della Torre E. *J. Appl. Phys.* 61, p. 4016, 1987.
48. Kadar G., Della Torre E. *IEEE Trans. Magnetics* 23, p. 2820, 1987.
49. Bean C.P. *Phys. Rev. Lett.* 8, p. 250, 1962.
50. Bean C.P. *Rev. Mod. Phys.* 36, p. 31, 1964.
51. London. H. *Phys. Lett.* 6, p. 162, 1963.
52. Kim Y.B., Hempstead C.F., Strnad A.R. *Phys. Rev. Lett.* 9, p. 306, 1962.
53. Salling C., Schultz S. *IEEE Trans. Magnetics* 24, p. 2877, 1988.
54. Restorff J.B., Savage H.T., Clark A.E., Wum-Fogle M. *J. Appl. Phys.* 67, p. 5016, 1990.
55. Potter R.I., Beardsley I.A. *IEEE Trans. Magnetics* 16, p. 967.
56. Beardsley I.A. *IEEE Trans Magnetics* 22, p. 454, 1986.
57. Koehler T.R. *J. Appl. Phys.* 61, p. 1568, 1987.
58. Slonczewski J.C. Unpublished IBM Report, 1961.
59. Prutton M. *Thin Ferromagnetic Films*, Butterworth, Washington D.C., 1964.
60. Pinto M.A. *Phys. Rev. B* 38, p. 6824, 1988.
61. Friedman G. *J. Appl. Phys.* 67, p. 5361, 1990.
62. Gradshtein I.S. Ryzhik I.M. *Tables of Integrals, Series and Products*, Academic Press, New York, 1980.
63. Madox J. *Nature* 321, 1 May, p. 11, 1986.
64. Lubich C. *IMA J. Numer. Anal.* 7, p. 97, 1987.
65. Joshi A.W. *Elements of Group Theory for Physicists*, Wiley, New Delhi, 1982.
66. Vilenkin N.J. *Special Functions and the Theory of Group Representation*, American Mathematical Soc., Providence, Rhode Island, 1968.
67. Lyubarskii G.Y. *Application of Group Theory in Physics*, Pergamon Press, Oxford, 1960.
68. Street R., Wooley J.C. *Proc. Phys. Soc. A*, 62, p. 562, 1949.
69. Tobin V.M., Schultz S. *IEEE Trans. Magnetics* 24, p. 2880, 1988.
70. Rodé D., Bertram H.N., Fredkin D.R. *IEEE Trans. Magnetics* 23, p. 2224, 1984.
71. Mayergoyz I.D. *Iterative Methods for the Calculation of Electromagnetic Fields in Inhomogeneous Anisotropic and Nonlinear Media*, Naukova Dumka, Kiev, 1979.
72. Tozoni O.V., and Mayergoyz I.D. *Analysis of* 3-D *Electromagnetic Fields*, Technika, Kiev, 1974.
73. Mayergoyz I.D. *J. Appl. Phys.* 57, p. 3803, 1985.

74. Doong T., Mayergoyz I.D. *IEEE Trans. Magnetics* 21, p. 1853, 1985.
75. Mayergoyz I.D. *Phys. Rev. Lett.* 56, p. 1518, 1986.
76. Mayergoyz I.D. *IEEE Trans. Magnetics* 22, p. 603, 1986.
77. Mayergoyz I.D., Friedman G. *J. Appl. Phys.* 61, p. 4022, 1987.
78. Mayergoyz I.D., Friedman G. *J. Appl. Phys.* 61, p. 3910, 1987.
79. Mayergoyz I.D., Friedman G. *IEEE Trans. Magnetics* 23, p. 2638, 1987.
80. Mayergoyz I.D., Friedman G. *IEEE Trans. Magnetics* 24, p. 212, 1988.
81. Mayergoyz I.D. *J. Appl. Phys.* 63, p. 2995, 1988.
82. Mayergoyz I.D. *IEEE Trans. Magnetics* 25, p. 3925, 1989.
83. Mayergoyz I.D., Friedman G. *IEEE Trans. Magnetics* 25, p. 3934, 1989.
84. Mayergoyz I.D., Adly A.A., Friedman G. *J. Appl. Phys.* 67, p. 5373, 1990.
85. Mayergoyz I.D., Keim T. *J. Appl. Phys.* 67, p. 5466, 1990.

Index

Abel type integral equation, 175, 181, 191, 195, 198
Accomodation, 108
Anisotropy energy, 142
Astroid curve, 144

Bean (critical state) model, 114

Chebyshev polynomials, 174, 195
Classical Stoner-Wohlfarth vector model, 141
Comparison between vector Preisach models and S-W model, 201
Congruency property of the Bean model, 119
Congruency property of the generalized Preisach model, 93
Congruency property of the moving Preisach model, 69
Congruency property of the Preisach model, 15, 16, 17
Congruency property of the vector Preisach models, 158, 159, 160

Definition of vector Preisach models, 149, 150
Demagnetized state, 23
Deterministic input, 56
Dynamic isotropic vector Preisach models, 188
Dynamic Preisach models, 99

Elementary hysteresis operators, 2
Energy of interaction of magnetic moment, 142
Expected value, 56
Experimental determination of the Preisach function
  using first-order ascending curves, 21, 22
  using first-order descending curves, 17, 18
Experimental testing of the Preisach-type models, 129

Fading memory, 17
Fast Fourier transforms, 180
First-order reversal (transition) curves, 18
Fourier expansions, 180, 198
Frechet derivative of the Preisach hysteresis operator, 43

Generalized critical state model, 119
Generalized S-W model, 148
Generalized two-dimensional anisotropic vector Preisach models, 198
Generalized two-dimensional vector Preisach models, 194
Generalized three-dimensional vector Preisach models, 196
Geometric interpretation of the Preisach model, 7

Higher-order reversal (transition) curves, 24

High-order restricted Preisach models, 95
Hysteretic energy dissipation, 4, 44, 45, 46, 47, 48, 49, 82

Identification problem, 4
Identification problem for the anisotropic vector Preisach models, 179
Identification problem for the dynamic isotropic vector Preisach model
  three-dimensional, 193
  two-dimensional, 190
Identification problem for the generalized S-W model, 149
Identification problem for the isotropic vector Preisach models, 172
Identification problem for the moving Preisach models, 67
Identification problem for the nonlinear Preisach model, 73
Identification problem for the Preisach model with accomodation, 110
Identification problem for the restricted Preisach model, 89
Identification problem for the vector Preisach model, 155
Intrinisic limitations for the classical Preisach model, 64

Limiting triangle, 7

Magnetic definition of the Preisach model, 6
Magnetic moment, 5, 141
Magnetic particles (domains), 5, 141
Magnetostrictive hysteresis, 122
Main properties of the Preisach model, 12
Matching nonzero initial slopes of reversal curves, 29
Memory formation, 11, 12, 17
Moving Preisach model, 64, 65, 66, 67, 78

Necessary and sufficient conditions, 4, 24
Noise, 56
Noninteracting particles, 55
Nonlinear (input dependent) Preisach model, 70
Numerical implementation of the dynamic Preisach models, 105
Numerical implementation of the nonlinear Preisach model, 82
Numerical implementation of the Preisach model, 32, 33, 34, 35, 36
Numerical implementation of the Preisach model with accommodation, 112
Numerical implementation of the S-W model, 148
Numerical implementation of the two-dimensional generalized isotropic vector Preisach model, 194
Numerical implementation of the vector Preisach model, 156

Particle distributions, 27
Preisach (weight) function, 2, 62
Preisach function symmetry, 23
Preisach hysteresis operator, 2
Preisach model, 2, 65
Preisach model as a neural network, 17
Preisach model with accomodation, 108
Preisach model with stochastic input as a model for viscosity, 55
Probability density function, 57
Property of correlation between mutually orthogonal components of output and input, 169
Property of equal vertical chords for the nonlinear Preisach model, 71, 77

Reduction property of the vector Preisach model to the scalar Preisach model, 157
Reduction theorem for the nonlinear Preisach model, 79
Relaxation process, 103
Relaxation time, 101, 102, 190, 191, 197
Representation theorem for the generalized Preisach models, 93
Representation theorem for the nonlinear Preisach model, 77
Representation theorem for the Preisach model, 24
Reptation, 108
Restricted Preisach model, 87

Reversible components of hysteresis nonlinearities, 12
Rotational symmetry of vector Preisach models, 161

Second-order reversal (transition) curves, 74
Some basic properties of the vector Preisach models, 157
Spherical harmonics, 183
Staircase interface, 11
Statistical instability, 27
Stochastic process, 56
Stoner-Wohlfarth magnetic particles, 32, 141
S-W model as a generalized mathematical vector model of hysteresis, 148
Superconducting hysteresis and the Preisach-type models, 114
Superposition model, 140
Switching probabilities, 57

Theorem, 39, 49, 85, 151, 174, 182
Thermal activation type models for viscosity, 59
Three-dimensional vector Preisach model, 154, 155
Two-dimensional vector Preisach model, 154, 155
Two-input Preisach model, 123

Vector Preisach models, 141
Vector Preisach models driven by vector stochastic inputs, 199
Viscosity, 59
Viscosity phenomena in hysteretic systems, 4
Volterra equation of the second kind, 186
Volterra integral equations, 185

Wiping-out property of the Bean model, 119
Wiping-out property of the moving Preisach model, 69
Wiping-out property of the nonlinear Preisach model, 71
Wiping-out property of the Preisach model, 13, 14, 15
Wiping-out property of the restricted Preisach model, 92